TECTONIC GEOLOGY

TECTONIC GEOLOGY

Prof. P.S. Saklani
Department of Geology, University of Delhi

SSPH
SATISH SERIAL PUBLISHING HOUSE
403, Express Tower, Commercial Complex, Azadpur, Delhi - 110033 (INDIA)
Phone : 011-27672852 Fax : 91-11-27672046
E-mail : info@satishserial.com, hkjain1975@yahoo.com
Website : www.satishserial.com

Published by :

SATISH SERIAL PUBLISHING HOUSE
403, Express Tower, Commercial Complex, Azadpur, Delhi - 110033 (INDIA)
Phone : 011-27672852 Fax : 91-11-27672046
E-mail : info@satishserial.com, hkjain1975@yahoo.com

ISBN 81-89304-22-4

Printed at
SALASAR IMAGING SYSTEMS, DELHI-35

FOREWORD

The **Tectonic Geology** is intended for graduate students. Elements of tectonics are briefly described. Principles of Plate Tectonics and types of Plate Margins are dealt with in the book. What are the causes for plate movements and mid-oceanic ridges are also included. How the Himalaya was evolved in various stages of uplift and northward drift of India are some important aspects. The tectonic architecture, various structural zones and models of metamorphism are included in the book. The Himalaya is seismically active and the seismic gap in Garhwal and Kumaun is prone to Earthquakes. Minor structurers reflect the major structures and are of five phases. Petrotectonics and structural petrology are significant tools for crystal deformation in Garhwal Region. The Himalaya and Alps have many similarities and contrasts. Thrust tectonics with modern approaches gives a new insight about the tectonic forces which affected the Himalayan Orogen. Neotectonics is significant in the Himalaya which has been reflected by the Earthquakes of the restless Himalaya. I hope the book would be useful to the students.

P.S. SAKLANI

CONTENTS

1

Tectonics

In general, the establishment of various types of structures of the crust is known as tectonics. In other words, regional setup of deformed features are constructed due to their interrelationship and origin. Some geoscientists do not consider much difference between structure and tectonics. In fact, the tectonics has a regional setting. Therefore, it can be said that structural geology primarily includes the geometrical elements (factors) while the origin of the structures and associated forces come under the purview of tectonics.

The development of major structures are commonly seen in folded mountain chains and metamorphic domains of nappe regions. The origin of those major structures in space and time depends to a greater extent on the types of sediments, igneous intrusions, processes of metamorphism and stresses. The setting of structural limits on the continents and oceans is also an important aspect of tectonics. Suess described about these factors in his book 'Das Anlitz de Erde (1883-1888 in Hills, 1973) which deals about the origin of major structures and their regional strike based on geological cross-sections. About origin and geological history of the major structures our knowledge is still not complete. To find the authentic data we are largely dependant on geophysical instruments and experimental results. It is thus clear that the tectonics is still in the progressive stage. In tectonic analysis, the movement of stresses in tectonic units of the crust and their relation to each other are comparatively studied.

The structures are controlled by the effects of stresses which can give rise to mountain building processes on a regional scale. Mapping of geological features are helpful to get an idea of tectonics.

OROGENY

A continent generally consists of mountains, plateaus and plains. It has also been seen that due to impact of compressional stresses the rocks of the mountain are folded, faulted and thrust while in the regions of the plateaus and plains the rock segments are less disturbed. The stresses responsible for rock-structures of the mountains cause the orogenesis and common structures are folds, faults, nappes and thrusts e.g. young mountains of the Himalayan and Western Alps (Fig. 1.1). Due to the effects of stresses the rocks of the mountains get subsided or uplifted. So the exact definition of orogeny is quite difficult. It is certain that in orogeny structures are developed on a regional scale in the sedimentary, metamorphic and igneous rock sequences.

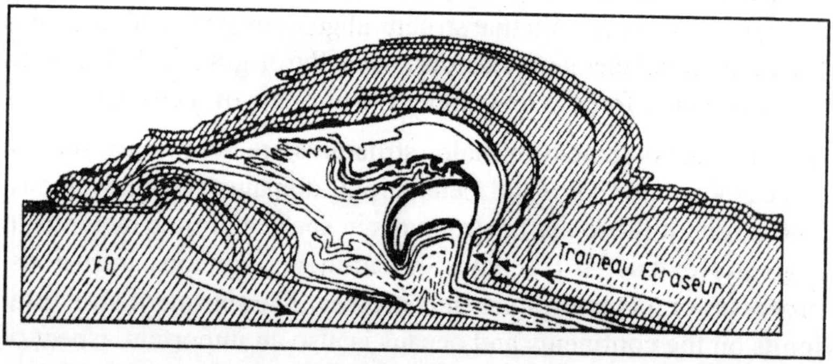

Figure 1.1: The nappe structures of the Western Alps, showping the Traineau ecraseur pushing against the alpine edifice and over-riding the foreland of Europe (Fo). *(After Argand, in Hills, 1973).*

EPEIROGENY

In contrast to orogeny, the epeirogeny is mainly characterized by vertically directed stresses which are responsible for the rise and subsidence of rocks. These stresses can give rise to formation of geosynclines and geanticlines. Continental forces do not form a special folded or faulted region. The depressions of geanticlinal regions are marked by continuous sedimentation for many geological periods.

TAPHROGENY

The tectonic setup of mountains and plateau areas, when is subjected to block faulting, results into the formation of graben and rift valleys and these movements are taphrogenic. The rock displacement in these regions is vertical rather than lateral during the orogeny. Monoclinal warping of rocks is also quite common in the areas affected by taphrogeny. The Narmada, Sabarmati, Rhine rift valleys are examples of taphrogeny (Fig. 1.2 A, B, C).

When the rocks are displaced along the strike then lineaments can be formed. The movements are horizontal and the lineaments formed are tectonic. The rocks along the strike slip faults are, at times, accampanied by secondary folds.

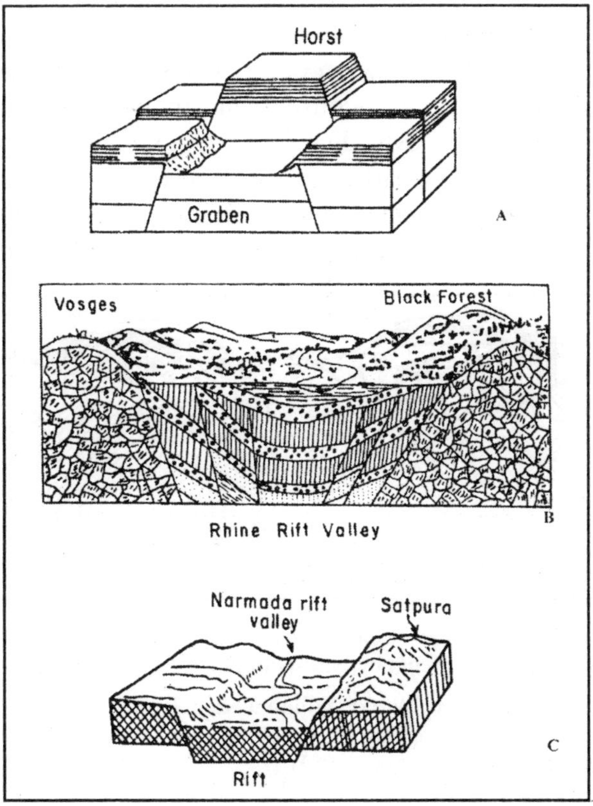

Figure 1.2: (A) Horst and Graben structures (after *Lahee,* 1971) (B) Rhine Rift Valley (after *Lange et.al.,* 1966) (C) Narmada Rift Valley.

Cratons, Shields and Geosynclines

Cratons are characterized by old, consolidated and folded rocks which are commonly affected by metamorphism.

The typical examples of cratons are the basement complex of Archaeans affected by later orogenies. The orogens of Palaeozoic and youner age are mobile, while the Precambrian folded belt are immobile and rigid and occur in domal plateau and are also known as shields (*e.g.* Canadian and Baltic shields *etc.*).

Geosynclinal concept was first proposed by *Hall* (1859 in Hills, 1973) and later by *Dana* (1893), *Kober*, (1955) *Stille* (1936), *Kay* (1947), *Auboin* (1959) and *Trumpy* (1960). These workers included all the sediment filled basins and troughs under gresynclines. The earlier stages of orogen are grosynclinal. Marine troughs unfilled by sediments differ from the sediment filled geosyncline. In a geosyncline the sedimentation occurs in a slowly sinking trough under shallow marine conditions and the sediments attain a thickness of 10000m extending for several kilometres (Fig. 1.3). Geosyncline formed in earlier stage of orogeny is orthogeosyncline associated with sedimentation and volcanic processes. The rocks are not associated with volcanic process in a miogeosycline.

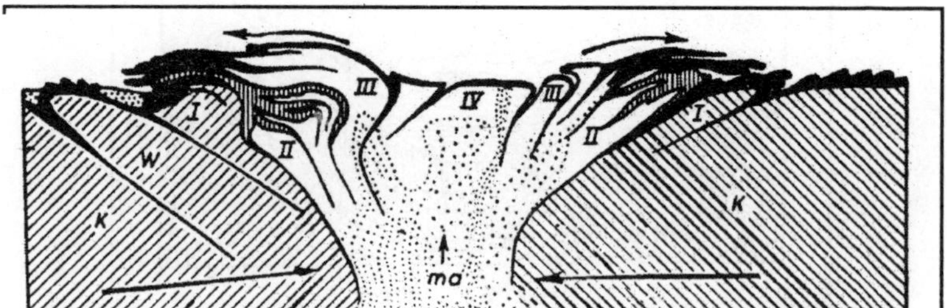

Figure 1.3: The structure of the Western Alps showing underthrusting of the geosynclinal sediments by basement blocks.

K, crystalline basement; W, crystalline wedge; I, autochthonous nappes; II, parautochthonous nappes; III, exotic nappes; IV, median massif. ma, granitic magma. (After Kober, 1955).

When the history of major structures is long and the geosynclinal sedimentation continues to be affected by local activity then the forces in general are vertical and characterized by various intrusions. These factors also bring changes in the sedimentation processes as is commonly observed in the mountain building. The end stage of major folding is marked by vertical displacement and the tectonic processes are renewed.

The structures of the folded mountains show similar orientation in tectonic zones and the rocks and grade of metamorphism are also similar. During orogeny commonly greywake and flysch rocks are formed. The setup of sedimentary facies is also changed due to nappe movments and the deformed rocks get inclined towards the foreland (craton) and its back parts are characterized by strong intensity of folding. The folds affected by compression are parallel to the axes of shear while in the medium type of shear the fold axes are horizontal and parallel. Shear folds show staircase orientation and the fold axes are at angle of 45° to the conjugate forces.

The folds developed at the basal parts of the rocks are different than the overlying sediments. The folds of the Jura mountain and those of the basal parts of the Alps are quite different. The basal parts can be affected to shortening and its main cause is thrusting.

The model envisaged by Kober (1955) for the Alps depicts underthrusting of the geosynclinal sediments by the flanking cratons when they were together. There is drag of convection current in the mantle at the base of the crust.

The shortining takes place due to downward trend of the layered granite.

The origin of tectonic forces can be explained by principles of plate tectonics discussed in the later pages. It is regarded that the Sial (Silica – Aluminium rich) granitic layer overlies the Sima (Silica – Manganese rich) basaltic layer. The granitic layer rather floates over the basaltic layer due to buoyancy created by density contrast of the minerals. As a result the basal part of Sial is shortened due to convection current in the earth interior. When the granitic layer goes down then the sediments of the basin get buckled in the

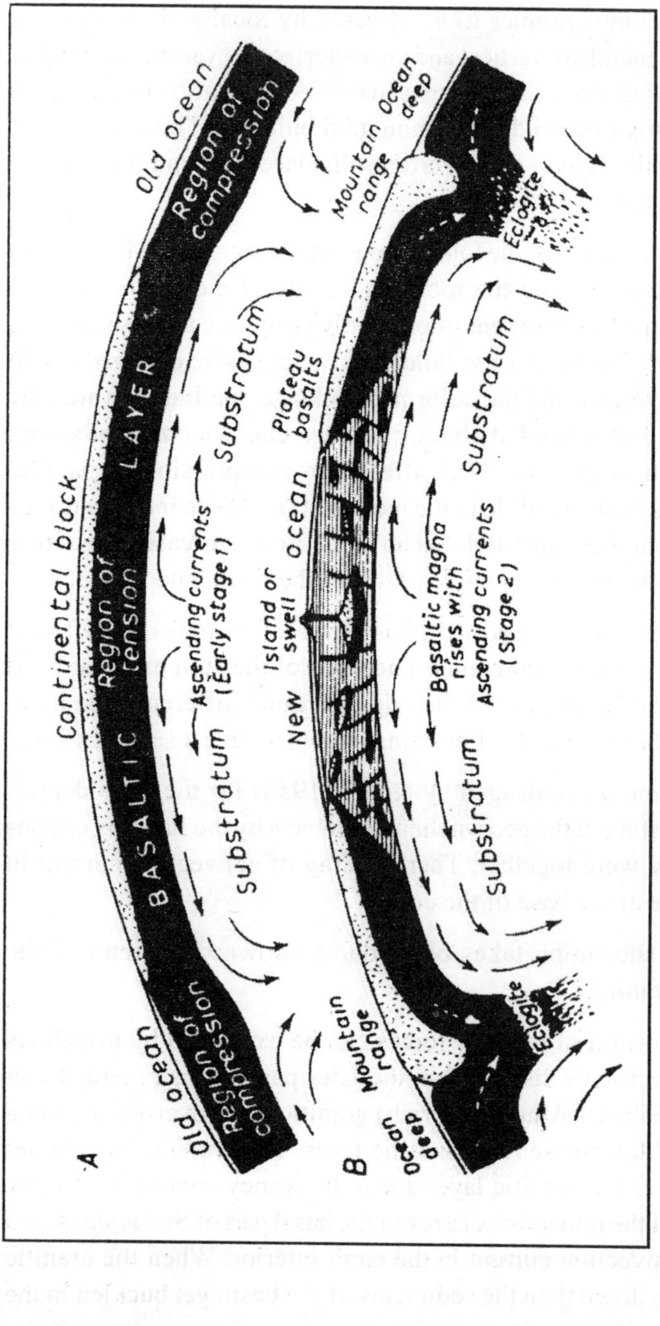

Figure 1.4: Folding due to convection currents in the Mantle (from *Holmes* 1965).

lower parts due to load of the overlying sediments. This down buckling gives rise to complex fold styles in sediments and the overlying sediments are subjected to thrusting associated with nappes. Such a type of displacement when is produced by vertical forces then slip is associated with gravity.

Gravitational Tectonics

Rock masses glide along a gentle slope which originates due to the activity of hypogene forces and are frequently associated with folding and faulting which range in size from minor to regional structures. Slumping and landsliding are not the products of gravitational tectonics because geological structures are associated with the creep of rocks and not disruption as is seen in slumps and landslides. The concept of gravitational tectonics was propounded by Lamont (1794), Kuhn (1836) and Naumann (1819-1850 in Hills, 1973 (Fig. 1.5).

Haarmann (1930) put forward the oscillation theory which explains that the orogenic structures can be formed by free gliding

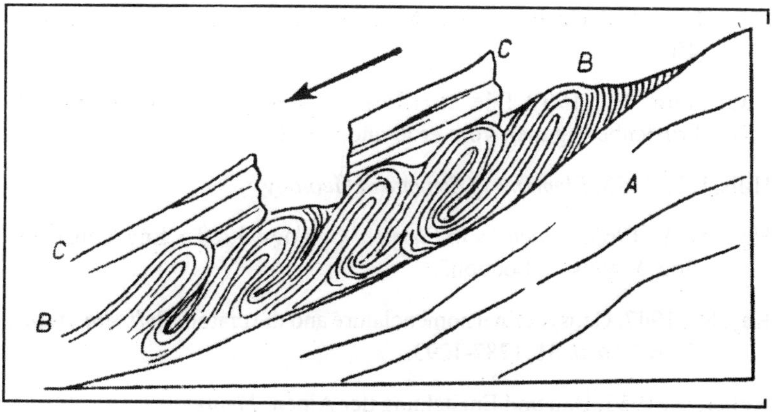

Figure 1.5: Gravitational Tectonics, according to Naumann (1850 in Hills, 1973).

under the influence of gravity or by gravitational sliding in a sedimentary trough or by gliding of plastic rocks.

Van Bammelen (1954) opined that vertical movements in the crust are followed by secondary tectogenesis controlled by gravitational gliding. Predominance of soft rocks such as plastic clay, gypseous shale, rock-salt can flow under confining pressure and are more susceptible to gravitation gliding.

Resurgent Tectonics

Old Faults and thrusts can be rejuvenated by later stresses. The rejuvenation of old troughs and renewal of movements on faults are due to neotectonics described separately. The old strike-slip faults can be transformed to normal faults due to the effects of later orogeny.

References

Auboin, J., 1959. A Propos d'un Centenaire les aventures dela Notion de Geosynclinal *Rev. Geog. Phy. Geol. Dyn.* **2**(3), 135-188.

Bammelen, R.W.V., 1954. Mountain Building, 177p *Mortinus Neg hoff.* Hague.

Dana, J.D., 1893, On the origin of Mountains, *Am. Jour. Sci., Ser.* 3, 5, 347-450.

Haarmann, H., 1930 Die oszillations theory line erklarung der krustenbewegunen vom erde und mond.

Hills, E.S., 1973, *Elements of Structural Geology.*

Holmes, A., 1965, *Principles of Physical Geology*, 1288 p. Eng. Lan. Book Soc. & Nelson, London.

Kay, M., 1947, Geosynclinal nomenclature and the craton, *Bull. Am. Assoc. Petrol. Geol.* **31**, 1287-1293.

Kober, L., 1955, Bau und Entstehung der Alpen, *Vienna.*

Kuhn, K.A., 1836, Handbuch der Geognoise, Leipzig Vol. I and Vol. II, 850.

Lahee, F.H., 1971. *Field Geology* (6th Edition McGraw Hill Book Co., Tokyo, 926 p.

Lange, O., Ivanova, M. and Lebedeva, N., 1966. *General Geology,* Foreign Languages Publishing House, Moscow, 299 p.

Lamont, F.P.W., 1974, Observations Geologiques Sur le Gisement et al forme des replis successifs etc., *Jour. des Pines,* **59**, Aun. 7, 449-454.

Stille, H. 1936, Einfuhrung in den Bau Nordeamerikas, Bd. II, 77-97.

Trumpy R., 1960, Palaeotectonic evolution of the Central and Western Alps, *Bull. Geol. Soc. Am.,* **7**, 843-908.

• □□□ •

2

Plate Tectonics

Francis Bacon (1620, Antonio Snider, 1858; F.B. Taylor, 1910 in Hallam, 1973) and E. Wegener (1937) gave their concepts about the continental drift. Taylor developed this hypothesis from the works of Edward Suess and described that the mountain ranges are the result of oceanic depressions and the tangential stresses are directed towards the ocean form horsts. This phenomenon happens due to cooling and contraction of the earth. Whenever the resistance is lost the crust flows out in lobes forming the mountains and arcs. The notion of crustal creep from high to low latitudes in the northern hemisphere was envisaged by Taylor with reference to Greenland, a ramanant of an old massif, from which Canada and northern Europe were broken along the rifts. Wegener (1935) based his hypothesis on geophysical data and was struck by the similarily of the coastlines on either side of the Atlantic. He argued that the continents can move vertically as well as horizontally. The ancient gneissic plateau of Africa bears much similarity with Brazil. The early Palaeozoic (Caledonian) and late Palaeozoic mountain belts of Europe can be traced in Newfound land and Nova scotia. The Gondwana flora of India and Australia can be correlated with each other. The tillitic rocks are present in S. Africa, S. America, India and Australia. Wegener (1937) emphasized on the changing path of polar wandering from Carboniferous, Eocene to lower Quaternary times. In upper Carboniferous time India was joined with East African Coast and was characterized by a northern drift. It continued in Eocene and its present position was attained in the lower Quaternary (Fig. 2.1).

The paper by Wilson (1957, 1963) ascribed to the formation of transform faults due to the movement of plates and its full theoretical

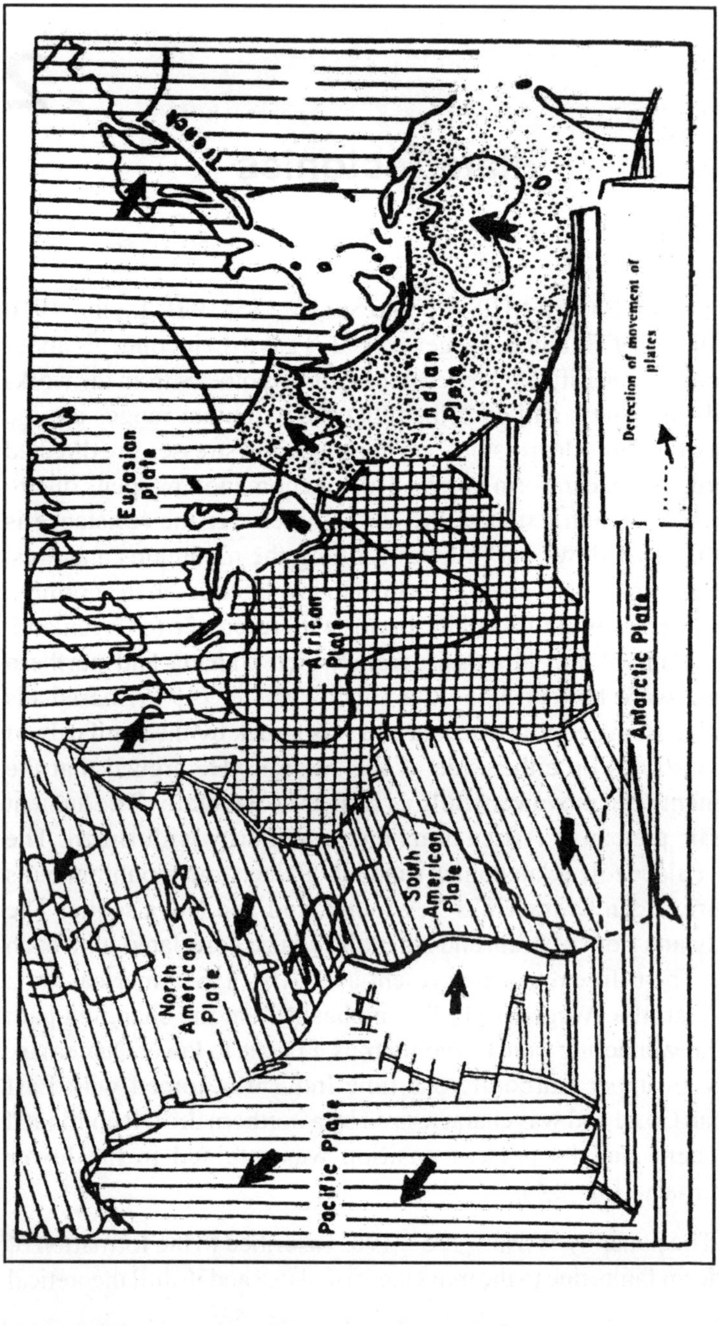

Figure 2.1: Major plates of the world.

formulation was done by Morgan, Mekenzie and Pichon (in Hallam, 1973). According to Morgan the earth's surface can be divided into 20 blocks associated with three types of boundaries. (1) Ocean rise where crust is created, (2) trenches where crust is destroyed and (3) transform faults where the crust is neither created nor destroyed. The blocks or plates are considered as rigid which extend upto a depth of about 100 km down the low velocity layer of the mantle. He referred this relatively rigid zone extending upto 100 km as tectonosphere.

The original approach in understanding the earth surface stimulated Pichon to attempt an ambitious survey of magnetic anomaly and fracture-data over the globe and suggested for the existence of six plates, namely, (1) American, (2) Eurasian, (3) African, (4) Indian, (5) Pacific and (6) Antarctic plates (Fig. 2.1). Le Pichon used two independent sets of data to determine the centers of rotation the spreading rates inferred from magnetic anamolies and azimuths of tranform faults at the intersection of oceanic ridge axes. Possibly, S. Pacific, N. Pacific, Arctic Atlantic and Indian oceans opened due to a single rotation.

The earth's crust and the upper mantle down to a depth of about 100 km (somewhat more beneath the continents and less beneath the oceans) is called as asthenosphere. The lithosphere overlies a zone of much weaker and hotter rocks and is marked by significant reduction is shear wave velocity. The lithosphere is a type of hard outer rind underlain by more plastic substratum.

The tectonic activity within the lithosphere has long been known to be largely confined to a series of narrow zones marked by earthquakes and volcanic activity, most notably, the "Firy Ring" around the margins of the Pacific.

These tectonically active zones can be referred to as boundaries of six major and twelve minor rigid plates. The plate margins are of three types :

(1) Constructive Plate Boundary

In this type the new crust is created by upwelling of material from the mantle signified by oceanic ridges. A brief description about the ocean and its features is necessary to understand the constructive plate margins.

An ocean can be divided tophographically into the following major provinces.

Continental margin consists of continental slope, shelf and deep trenches. Continental shelf partly consists of fault structures and partly by shallow water sediments, pelagic deposits, red clay / forminiferal-ooze in the core and sands carried by turbidity currents. Sea mounts rising from the sea floor are volcanic (basaltic) in origin. In the west of Pacific ocean these are flat topped and coral deposits which occur on the top of seamounts indicating shallow water conditions of deposition.

Midoceanic ridge is characterized by rugged topography and block faulting. The ridge occurs in the form of a swell and is of regional extent. The swell can be for more 1000 km in width (*e.g.* East Pacific rise). Its arial zone is zone of shallow focus earthquakes. Its relief in some areas is a kind of rift valley. The Mid Atlantic ridge measures 25 to 50 km in width and is deep from 250 to 750 meters (Fig. 2.2).

The East Pacific Rise enters indo the gulf of California and mountains of Western U.S.A.

The common rocks of the midoceanic ridge are basalts gabbros and amphibolites.

Heat flow under the axes of midoceanic ridges is found to be much higher than elsewhere in the ocean.

As remarked earlier, the midoceanic ridges are associated with faults.

The San Andreas fault of California from San Francisco to east of Los Angeles is dextral and is regarded to shift northwards at the rate 5 cm per year.

The Alpine fault (transcurrent) in New Zealand is parallel to the margin of the Pacific coast. The Great Glen Fault in Scotland is a sinistral transcurrent fault. The faults in which the dominant movement of the block against the other is lateral rather than vertical, are transcurrent megashear.

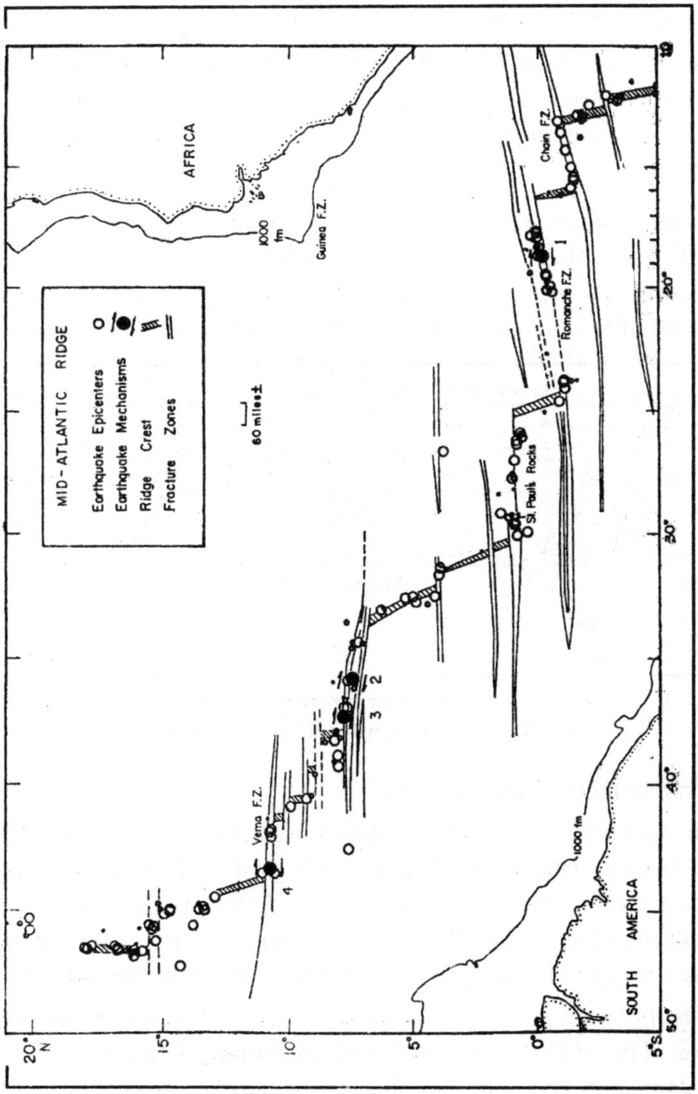

Figure 2.2: Transform faults on Mid-Atlantic Ridge. (After Sykes), in Billings, (3rd edition, 2000).

The faults in which the displacement is suddenly stopped or which change their orientation are termed as *transformed faults*.

(2) Destructive Plate Boundary

In such a type of plate boundaries the crust is consumed by being forced down beneath the island arc – ocean trench systems or mountain zones and the cold lithosphere descends down into the asthenosphere.

These curvilinear features are referred to as subduction or Benioff-Zones (Fig. 2.3). In fact, Hugo Benioff (1954, 1958), a seismologist, emphasized the importance of seismic zones indicative of faults inclined towards the continents.

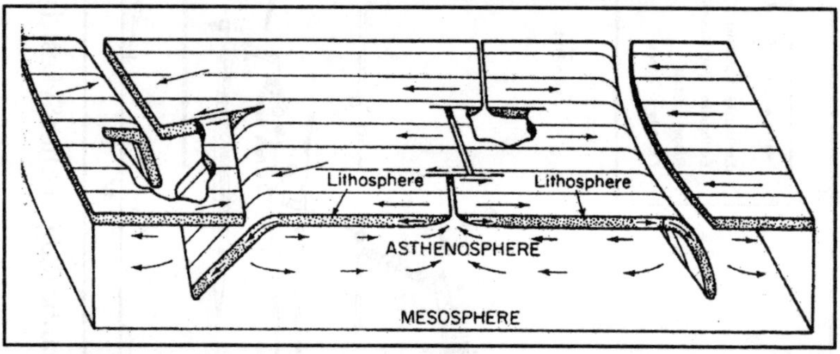

Figure 2.3: Plate tectonics and sea-floor spreading. The lithosphere plunges downward along the Benioff zone. (After Billings, 3rd edition, 2000).

(3) Conservative Plate Boundary

In such a type of plate boundaries neither the crust is created nor destroyed. The plates slide laterally past each other and are characterized by transform faults beyond the offset ridge axes defining the rotation circles for the previous motions and hence provide a key to the direction of plate movements in the past.

The meeting point of two or three plates is known as a triple junction. This is the only way that the boundary between the two rigid plates can end. The boundary between the lithosphere and asthenosphere is called the zone of decoupling. It is possible that all the plate boundaries move with reference to a fixed geographic axis but the amount of crust created by spreading must balance the amount destroyed by thrusting along the Benioff zones. The spreading away from the ridge axes can be estimated by the magnetic anomaly data.

There is an important break between lithosphere and asthenosphere. The continental crust because of its buoyancy can not be carried to depth. As a result of this when two continental plates collide at a destructive zone the thickening of the crust is a common phenomena and the best example of this is the Himalaya. Due to collision of Indian and Asian plates the Himalaya was formed and the crust nearly doubled in thickness at the site of the subduction zone.

Causes for the Movement of Plates

Out knowledge about the mantle is still not complete and hence the cause of plate movement would remain speculative for a considerable time.

Hess (1955) and Runcorn (1961, 1962) proposed as model of convection current with the migration of sea floor within the mantle being coupled directly to upper limbs of the underlying convection cells. One fails to account how the rising and descending limbs of convection cell (Meinesz, 1962) can abruptly off-set by transform faults. It is hard to imagine the motion rate of rising limb under the ridges increasing away from the axes of rotation corresponding to the plate movement.

Elasser (1950) proposed that cold descending slab of lithosphere beneath the island areas is a major source of gravitational energy. Density contrasts with the surrounding asthenosphere will cause sinking and the sea floor will be pulled away alongwith mantle material welling up to fill the gaps created by ridge axes (Dietz, 1961). Sinks can act as stress guides once they have been initiated but hypothesis does not account about their initation. It is also doubtful if the motion can be maintained for a long time. The median position of the oceanic ridge and symmetry of spreading is easier to account for by passive dyke-fill of the rifts rather than active intrusion forcing the ridge flanks apart. The elevation of the ridges compared to flanking plates can be a possible source of gravitational energy. Thus, plates might partly slide under the gravity but it is not a major factor for plate tectonics.

Therefore, thermal convection of some kind appears to be a sufficient source of energy. It also applies to Elasser's concept in which the descending slab is seen driven by a temperature induced density contrast between it and the hotter mantle material through which it sinks.

E.R. Oxburgh, D.L. Turcotte and D.P. Mckenzie (in Hallam, 1973) also proposed possible convection models based on mathematical considerations. They envisaged a type of slow creep in crystalline rocks and its rate of flow is enhanced by a high ratio of actual temperature with that of melting temperature associated with low stress-differences. If temperature ratio is low or the stress difference is high then the rocks instead of deformation will be fractured.

Heat is transferred by a kind of thermal conveyer belt including the ascending hot and descending cold plumes. The slab eventually undergoes a type of thermal erosion as it is warmed by the conduction from the surrounding rocks. Although the lithosphere is supposed to be cooled by the time it commences to descend, island arcs are marked by high heat flow and abundant volcanicity. The necessary heat for the rock-fusion is provided by friction as the slab descending in the Benioff Zone.

Morgan (in Hallam, 1973) described about the hot spots of the mantle as evidences for volcanic island chains (*e.g.* Hawaiian islands). These islands progressively older from east to west, are regarded to be volcanoes erupted from a hot-spot in a fixed position with the westward spreading ocean floor. It is assumed that such rising hot spots (plumes) of mantle material start below the asthenosphere. Horizontal currents in the asthenosphere radiating from the plates form a 'thunder head' pattern of convection and can sufficiently account for the movement of plates. The actual pattern of motions is determined by the position and orientation of the plate boundaries.

References

Benioff, H., 1954. Orogenic and Deep Crustal Structure - Additional evidence for seismology, *Bull. Geol. Soc. Am.,* **66,** 385-400.

Benioff, H., 1958. Circum Pacific Tectonics, *Pub. Dom. Observatory, Ottawa,* **20**(27), 395-400.

Billings, M.P., 2000. *Structural Geology,* 3rd Ed., Prentice Hall, New Delhi, 606 p.

Dietz, R.S., 1961. continent and Ocean basin Evolution by spreading of the Ocean Floor, *Nature,* **190,** 854-885.

Elasser, W.M., 1950. The earths interior and geomagnetism. *Rev. Mod. Physics,* **22,** 1-35.

Hess, H.H., 1955. Serpentine, Orogeny and Epeirogeny, *Geol. Soc. Am., Sp. Paper,* **62,** 391-407.

Hallam, A., 1973. *A revolution in the Earth Sciences from Continental drift to plate tectonics, clarendon Press, Oxford.*

Hills, E.S. 1973 *Elements of Strucural Geology* 483 p., Methuen & Co. Ltd., London

Meinesz, F.A.V., 1962, Pattern of Convection current in earth's mantle, *Proc. Kon. Ned. Aka Wet.,* **65,** 131-143.

Runcorn, S.K., 1961, Climatic change through geological time in the light of palaeomagnetic evidence for Polarwandering Continental drift. *Quart. Jour. Met. Soc.,* **27,** 282-313.

Runcorn, S.K., 1962. Palaeomagnetic evidence for continental drift and its geophysical cause. In : *Continental Drift* - Int. Geop. Ser., **3,** 1-40.

Wegener, A., 1937. La Genese des Continenets 236 p., Nizet et Bastard, Paris. (Translation of German edition of 1936).

Wilson, J.T., 1957. Origin of Earth Crust, *Nature,* **179,** 1-48.

Wilson, J.T., 1963. Evidence from Islands on the spreading of Ocean Floors, *Nature* **197,** 536-538.

Wilson J.T., 1963. Hypothesis of Earth's behaviour, *Nature,* **198,** 925-929.

• □□□ •

3

Himalayan Origin

The Himalaya, through the ages, has been the maker as well as the defender of Indian and Trans-Indian civilization. For scientists, particularly the geoscientists, it has preserved an immense wealth of geological treasures which have been explored partly by successive generations of painstaking researchers. The Himalaya is really a thing of worship, as it has inspired the scientists, mountain adventurers and a vast number of laymen and spiritualists. It infact, is a huge scientific laboratory which they can use for ages.

One could be wonderstruck to know that this highest mountain chain evolved from a deep basin or a sea depression. The birth and development history of the Himalaya is of paramount interest but at the same time its evolution is highly controversial.

The classical view about the emergence of the Himalaya is that it is a result of the southward movement of the Asian continent which collided with the stable Aravalli-Vindhyan peninsular block. The phenomenon is further evidenced and supported by the precipitous slopes having a convexity towards the Indian side of the Himalaya. Contrary to this, the Trans-Himalayan parts in Tibet are gently sloping.

The world's highest mountain peak, the Everest, is situated in the Himalaya. It also has the honour of owning several mountain summits which excel in height located elsewhere on the globe. However, it is indicative of the fact that the vertically directed forces played a significant role in the later phases of the Himalayan configuration which had occurred about one million years ago.

It may be remarked that the formation of the Himalaya did not take place all of a sudden, rather its development was gradual.

Its formation was initiated about 310 million years ago and this age is known as Permo-carboniferous belonging to Palaeozoic era.

As remarked earlier due to collision of Indian and Tibetan plates the Himalaya was formed. India was a part of Gondwana. The collision and subduction between between India and Tibet occurred at equatorial latitudes and the suturing of India began at about 55 m.y. (Naeger et al., 1989; Klootwijk et al., 1992 in Burg., 2006). This displacement rate from 15-25 cm per year was decreased to 4-5 cm per year and the shortening rate is presently 1-2 cm per year (Demets et al., 1990 in Burg, 2006). Patriat and Achache (1984) described about the movement of India and Eurasia which occurred between 61 and 59 m.y. (Figure 3.1).

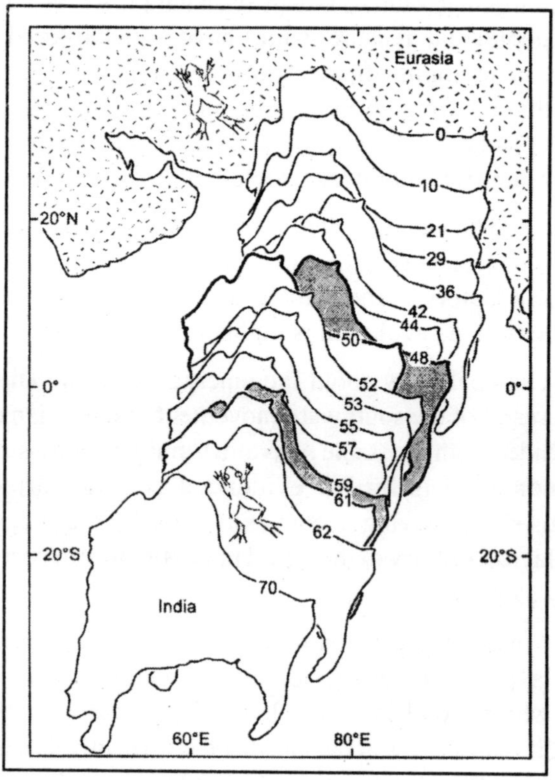

Figure 3.1: Northward drift of India with respect to Eurasia since 70 Ma (after Patriat and Achache, 1984). Numbers on the northern boundary of India indicate time in Ma for that position. Frogs refer to 65 Ma old continental bridging between India and Asia.

The river Ganga also originated alongwith the development of the Himalaya. Geoscientists believe that in the Palaeogeographic times the Himalayan sea or basin was squeezed into staggering large river known as the Indo-Brahm river. It was located between the Himalaya and the Aravalli-Vindhyan Peninsula. It is also called as the Siwalik river. In due course of time this river disappeared and was divided into Indus, Ganga and Brahmputra rivers flowing in different directions from one another. The disappearance of this very ancient river was probably caused by the formation of Kashmir and Assam syntaxial hair-pin like bends. This activity in the Himalayan region took place for more than 60 million years ago.

The second uplift of the Himalayan sea or basin took place in late Eocene times, i.e., about 50-60 million years ago. The main rocks formed during the periods constitute presently the Kirthar, Subathu, Jaintia, Dagshai, Kasauli and Barail hills.

The third but most violent phase of the emergence of the Himalaya belongs to middle Miocene times, i.e. about 25-30 million years ago. During the period the Himalayan sea/ depression rose to form lofty mountain ranges. Besides, the foredeep towards the southern side of the Himalaya was formed. This had witnessed prolific sedimentation thereby causing the formation of Siwalik rocks. The life during the Siwalik time was full of mammalian animals. However, man was not born upto this time. The climate during the Siwalik time was warm and humid and the landforms were clothed by thick forests. During the third event the layered rocks of the Himalayan depression were buckled and fractured. The large mountain blocks moved along the major factures for several kilometres. Due to this fact, it is said that the hills used to fly in the very ancient past. Most of the rocks did not remain at their original place and their order of disposition was also disturbed. Older rocks piled over the younger ones and made the structural architecture highly complicated (Fig. 3.2).

The Older rocks have travelled for about 110 km from their source and are also called as the root-zone. Important economic minerals have been found to be concentrated along the root zone; for example, the Ural mountains in Russia. The Himalayan root

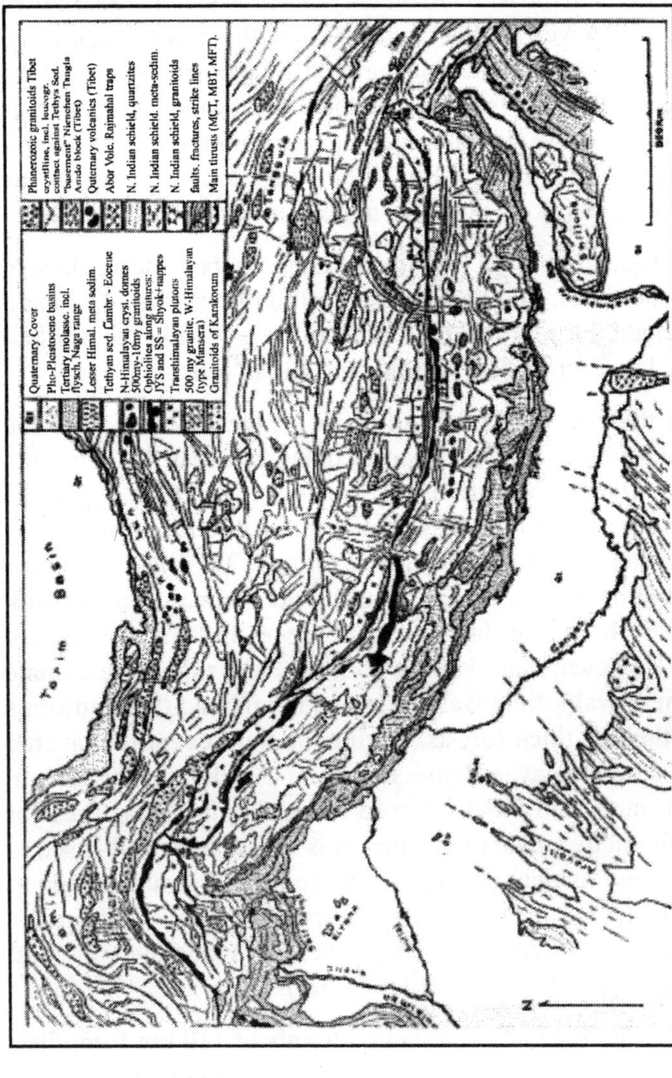

Figure 3.2: Structural map of the Himalaya and the greater part of Tibet. Compiled from various sources including Landsat photos by A. Gansser, 1993. A=Ambo, I+Islamabad, K=Katmandu, L=Leh, Lh-Lhasa, T=Thimpu, CL=Chomo Lungma (Everest), K2=(Chogori) with Karakorum fault zone, NB=Namche Barwa, NP-Nanga Parbat.

and other sub-parellel root zones preserve magmatic minerals such as copper, lead, zinc, silver, and antimony, topaz, sapphire and many gem stones occur in rocks of the root zone (Indus Sutre Zone).

The rocks belonging to the Tal are characterized by potential reserves of phosphorite, mainly used as a fertilizer. The Tal rocks are exposed at Maldeota near Mussoorie. Copper occurrences have been reported from the Krol rocks similar in age with Tal. Uranium minerals have been found in Ghansyali area of Garhwal.

The fourth event of the Himalayan orogeny took place in Pliocene time, i.e., about 12 million years ago. The warm and humid climate of the Miocene time suddenly became cold due to intervention of some natural forces including geological processes and the land forms were swept over by the large-scale glaciation. As a result, most of the animals of the Siwalik ranges were buried and died under the snow. The fossilized skeletons of these animals occur in rocks of the Siwalik hills.

The fifth and the last uplift of the Himalaya came into existence in Pleistocene time, i.e. about one million years ago and probably vertical forces played a significant role in this orogenic development of the Himalaya (Wadia, 1957). The Karewas of Kashmir reveal very interesting features concerning to this event. The Karewas contain some fossilized leaf impressions resembling very much the presently occurring Quercus (oak). The oak tree grows at an altitude of about 2000 metres but the fossilized leaf-impressions of this occur in the Karewas at an altitude of about 4000 metres indicating thereby the vertical uplift of the Karewas for about 2000 metres.

Possibly, human being appeared on this earth during the Pleistocene time and witnessed this geological deluge of the Himalayan turmoil.

I had personally done some field work in the Garhwal region and the geological data gathered therein indicate that the Himalayan rocks were structurally deformed by horizontal forces which formed large-scale buckles or folds and major fractures (also called as faults or thrusts). The vertical forces also brought significant structural changes in rocks of the Himalaya.

In fact, the Himalaya extends from Nanga Parbat in Kashmir to Namcha Barwa in Assam and the scientific investigations are of restricted nature. Therefore, this huge mountain belt should be explored with integrated efforts and then only the more conclusive picture of the Himalaya would be known from the point of view of its hidden treasures.

References

Burg, J.P., 2006. Two orogenic systems in the Himalaya : evidence and consequences., In : Saklani, P.S. (Ed.), *Himalaya (Geological Aspects),* **4**, Satish Serial Pub. House, Delhi.

Gansser, A., 1993. Facts and theories of the Himalaya. *Jour. Geol. Soc. Ind.,* **41**, 487-508.

Patriat, P. and Achache, J., 1984. India-Eurasia Collision chronology has implications for crustal shortening and driving mechanism of plates. *Nature,* **311**, 615-621.

Wadia, D.N., 1957. *Geology of India,* (3rd edition), Macmilan, London.

•□□□•

4

Himalayan Tectonics

The Himalayan orogen measures about 2400 km in length with a width of about 230 to 320 km with an average of 270 km. In the opinion of Gansser (1964, 1993) the rocks of the Himalaya represent entirely of shield material and platform sediments except for a narrow zone of ophiolitic belt. The Himalaya consists of major six zones from north to south. The northern zone i.e. *Zone 1* is composed of ophiolites associated with radiolarites, flysch and ultrabasics affected by thrusting. It, in order of abundance, constituted by sedimentary, plutonic, volcanic and metamorphic rocks. The plutonic rocks are ultrabasic sheets and exotic blocks contain olistoliths in olistromal flysch. It structurally consists of compressional root-folds and incompetent folds due to upper Cretaceous submarine gliding. The thrusting and gravity gliding occurred in Mio-Pliocene time (Fig. 4.1).

The rocks occur in a thrust sheet with a thickness of more than 300 m and include serpentine, radiolarite, limestone, dolomite, shale and sandstone. The peridotite and diabase were emplaced during their deposition. Exotic blocks were emplaced in the flysch during upper Cretaceous. All the rocks were thrust northward over the Tertiary molasses. The rocks are folded, sheared and boudinaged and step faults are also present (Gansser, 1964). The *Zone 2*, in order of abundance consists of sedimentary, plutonic and volcanic rocks. The Tibetan Himalaya (*Zone 3*) is a broad thrust belt mainly of Palaeozoic-Mesozoic sediments associated with plutonic rocks. The granitic intrusions cut the folded Mesozoics. The age of Everest granite of the Higher Himalaya is 15-16 m.y. (Krummenacher et al., 1978) and the contact metamorphic effects are also present. Precambrian and Cambrian strata are phyllitic. Ordovician strata

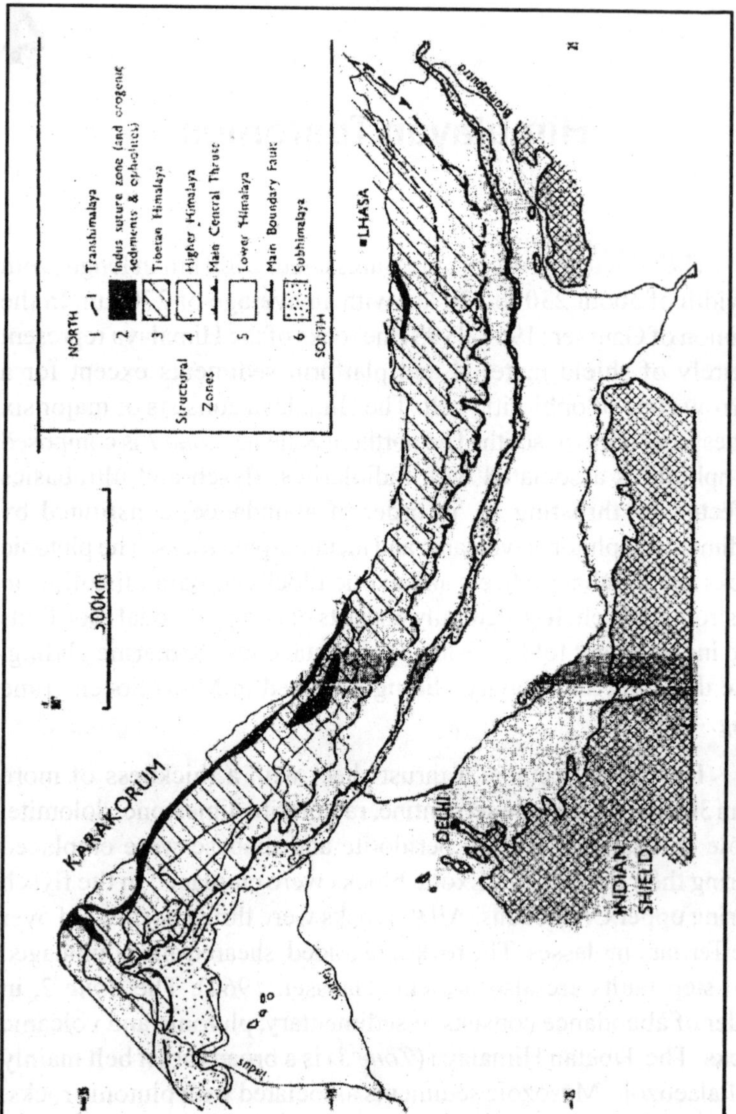

Figure 4.1: Structural zone of the Himalaya (After Gansser, 1964).

are nonmetamorphic. Folds and small thrusts developed during late Cretaceous to Pliocene. The Higher Himalaya (*Zone 4*) is a vertically uplifted region where the pre-existing sediments are eroded and consist of pre-Ordovician metamorphics intruded by granites. It consists of predominantly metamorphics with subordinate sedimentaries. The lower Palaeozoic sediments overlie the metamorphic rocks (Precambrian) and are intruded by Neogene granites. The high grade metamorphics occur at the base overlying the less metamorphosed to unmemorphosed lower Palaeozoics. Thursting movement affected the rocks associated with folds (isoclinal).

Further south of it, is the Lower or Lesser Himalaya (*Zone 5*) consisting of rocks of Precmbrian to Mesozoic age which are confined into nappes. It is sharply delinated by the Main Central Thrust (MCT) in north and Main Boundary Thrust (MBT) in south. It, in order of abundance, consists of metamorphic, plutonic and sedimentary rocks. It is characterised by nappes. Pre-Cambrian metamorphism and deformation affected the rocks but Palaeozoic strata remained unaffected. Its rocks show reverse metamoprism in thrust sheets. The southern most zone is the Sub or Outer Himalaya (*Zone 6*). It is located at the northern margin of Indo-Ganga plain and consists of fresh-water Siwalik sediments which are dissected by recumbent folds (southward vergence) and thrusts (Middle to late Pleistocene) and it is tectonically active.

The orogenic movements started in upper Cretaceous and affected the rocks of *Zone 1 and 2*. Subsequent to Indus suture zone, the intracrustal thrusting (Oligocene-Miocene) formed the Main Central Thrust (MCT). This thrust sheet contains the central crystallines measuring about 15 km in thickness with a cover of Tethyan sediments (12 km). The displacement of crystallines along this thrust is about 100-150 km towards south and the rocks are piled over by the late Precambrian and less metamorphosed sediments (Fig. 4.2). The MCT is an imbricated zone and gave rise to several nappes e.g., Almora, Jutogh, Chail, Budhakedar nappes etc.

According to Saklani (2005) the Main Central Thrust is represented by Vaikrita (MCT-1), Jutogh/Munsiari (MCT-II) and

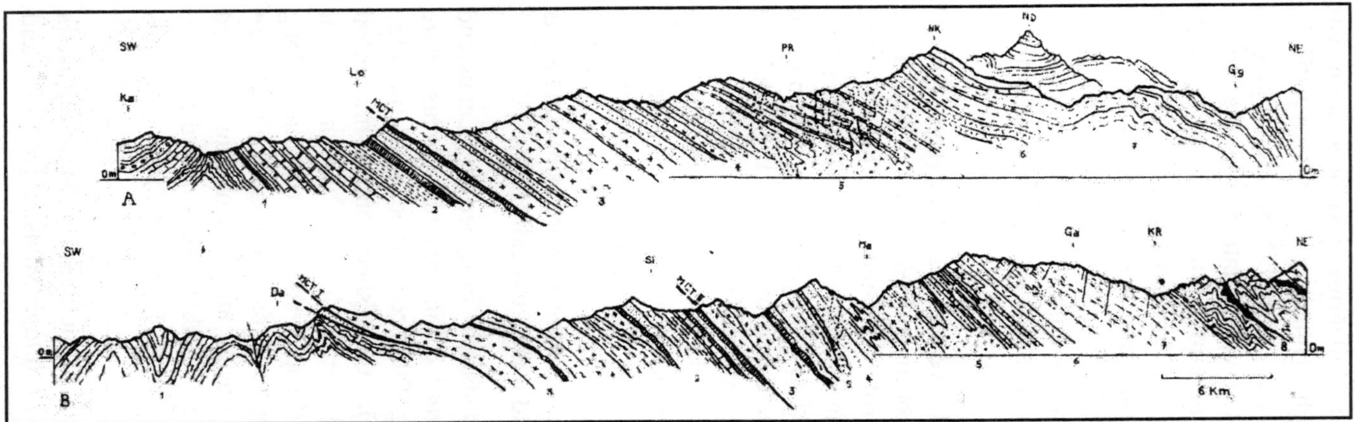

Figure 4.2: *Geological sections through the MCT and the Crystalline Core of the Kumaon Himalayas* (Central Himalayas). (After Gansser 1964, 1993). Section A through Nanda Devi. Section B along Kali River (W border of Nepal), to Garbyang, 1. Precambrian sediments of Lesser Himalaya. Metamorphism increasing towards MCT. 2. Pre-cambrian metasediments of Lesser Himalayas with schists, quartzites and amphibolites just below thrust, repeated in Sirdang area below MCT II.3. Gneisses, partly migmatitic of lower part of crystalline core. 4. Metaquartzites and high grade schists in middle section of crystalline. 5. Leucogranites with dykes intruding zone of marbles, limesilicates and psammite-gneisses. The relation of leucogranites with the upper high grade carbonate zone of the main crystalline is widespread.A certain three-fold division of the main crystalline core is also evident in Nepal and the eastern Himalayas. 6. Typical biotite porphyroblast-schists, ending the main metamorphism of the crystalline core (Budhi schists). 7. Late Precambrian to Cambrian argillaceous Tethys-Himalayan sediments. 8. Folded and imbricated sediments from Ordovician to Permain of the Tethys Himalays. Da = Darchula, Ga = Garbyang, Gg = Goriganga (Valley), Ka = Kapkot, KR = Kuti Rivar, Lo = Loharkhet, Ma = Malpa, ND = Nanda Devi (7,820m), NK = Nanda Kot, PR = Pindari River, Si = Sirdang. (After Gansser, 1993).

the Chail/Budhakedar (MCT III) in the form of a major duplex.The Central Crystallines and metamorphics showing high, medium and low grade of metamorphism are confined into the nappes along these thrusts. The rocks were remetamorphosed from Eocene to Pleistocene. The rocks were affected by crustal shortening of about 500 km. Folding and thrusting movements continued and affected the molassic southern edges of the Himalaya. The relics of the Gondwana directly cover Precambrian basement rocks. The Siwaliks were asymmetrically folded and affected by superficial thrusts which become bedding plane thrusts at depth with a shortening of about 5 to 10 km.

The Lower Himalayan structure is quite complex. It is thrust over the Siwaliks along the Main Boundary Thrust. It is sub-vertical on the surface. The rock units occur in imbricated stack and secondary branch line thrusts. The Gondwanas of this tectonic zone occasionally contain fossils. In the Eartern Himalayan part the Daling sediments and the crystallines are thrust over the Gondwanas and Eocene rocks (e.g. Shali window of Himachcal Pradesh and Ranjit window of Darjeeling area). The thrusting took place for about 30 km over the Siwaliks and was for 100 km in the Lower Himalayan region of Precambrian sediments.

MODELS OF FAULTING AND METAMORPHISM

In recent years, as many as four models have been proposed for the reversed metamorphism, anatexis and faulting within the Himalaya (Fig. 4.3).

The first model of Le Fort (1975) depicts a casual relationship between anatexis within the Higher crystallines and reversed metamorphism. It was proposed that the thrusting along the MCT heated the footwall with associated dehydration reactions. These fluids in the hot hangingwall fluxed the gneisses forming the leucogranitic melt and the heating was reponsible for the reversed metamorphism.

The lower Himalaya is characterized by the peculiar reversal of metamorphism towards the MCT and above of it. Some geologists still regard it due to large recumbent folds forming the nappes. This possible cause was suggested by Loczy (1907) in the

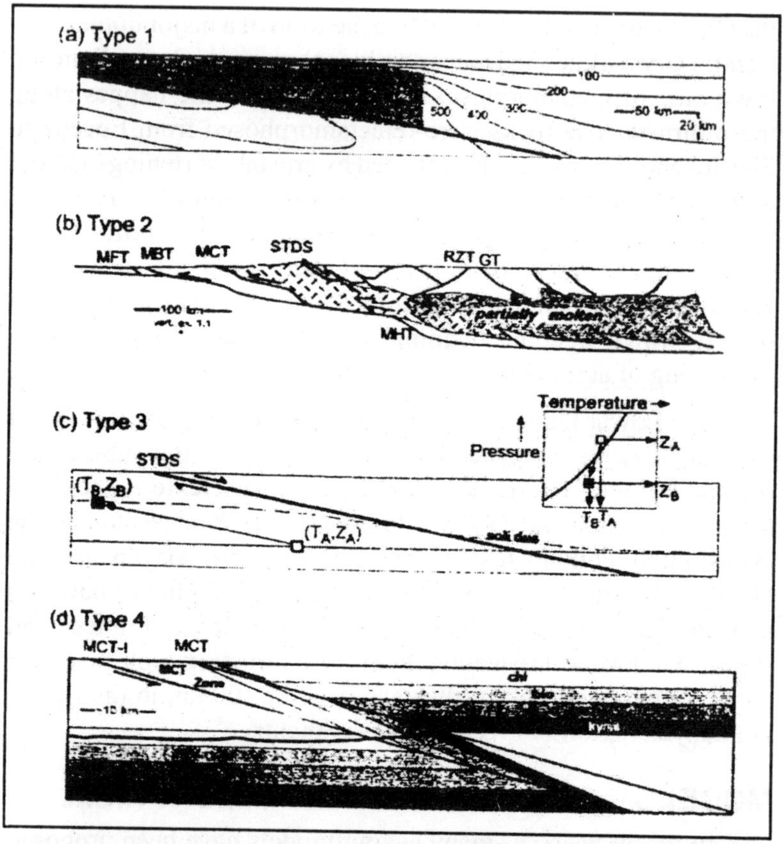

Figure 4.3: Schematic illustrations of the four different models proposed to explain the relationship of inverted metamorphism and or anatexis to large-scale faulting witin the Himalaya. (a) Type 1: Inverted metamorphism developed within the footwall of the MCT and anatexis in the hanging wall are spatially and temporally related by thrusting. The direction of material transport is shown by the line with the arrowhead, and the shaded region corresponds to crust accreted to the hanging wall. (b) Type 2: Thrusting results from anatexis. The illustration shows a partially molten Tibetan middle crust that is being actively extruded southward. In this interpretation, the Greater HimalayanCrystallines is an earlier extruded equivalent. (c) Type 3: Anatexis results from normal faulting. This figure illustrates the change in the depth-temperature conditions in the footwall of a normal fault. The initial depth. Za, is associated with temperature Ta. Following slip on the normal fault, the depth is substantially reduced to Zb, but the effect of heat results in only minor cooling to temperature Tb. (d) Type 4: Apparent inverted metamorphism in the footwall of the MCT is produced by deformation of two right-way-up metamorphic sequences. This illustration, modified from Hubbard (1996), shows the effect of deforming a normal metamorphic sequence across a broad shear zone. (After Harrison et al.,1999)

metamorphics of Sikkim where low grade chlorite schists are followed upwards by sillimanite gneisses (Heim and Gansser, 1939; Pecher, 1978; Pecher and Le Fort, 1986; and Hodges et al., 1988). Gansser (1993) remarked that the delicate dinocysts show their preserved membrane located below a 15 km thick crystalline slab and questioned the authenticity of the hot slab over the cold slab conept. As far as the MCT is concerned it shows a sharp contact in Eastern Himalaya and is imbricated in Central Himalaya while it shows its migration into the Himalayn foot-hills.

Auden (1935) ascribed about the reflected Aravalli trends in the Himalaya affecting the structures of the Himalayan rocks. The NNE directed Aravalli trends are extensions of the northern part of Indian shield. The N-S directed faults rejuvenated the Himalayan fracture system (Valdiya, 1976, 1980, 1981; Gansser; 1993).

According to the second model the thrusting in the Himalaya was caused by heating / melting. Bird (1978) suggested that continental collision caused lithospheric delamination accompanied by crustal uplift and this became the site for thrusting. This phenomena generated sufficient heat for anatexis and reversed metamorphism during 5-10 m.y. Davidson et al. (1997) described that syntectonic migration of leucogranites at the basal part of the Higher Himalayan crystallines which caused metamorphism within the hangingwall of the MCT.

The Third model relates to anatexis, melting and associated slip of the Southern Tibetan Detachment. (Harris et al., 1993).

The fourth model (Hubbard, 1996) is based on the fact that continuous convergence was associated with episodic magmatism and out of sequence thrusting was associated by right way up metamorphic sequences which were responsible for reversed metamorphism (Harrison et al., 1999). They (1998) also opined that the reversed metamorphism was created in the footwall and was later juxtaposed against the granite source region in the hangingwall.

The structural and metamorphic events of various phases ended with intrusions of leucograntic cuttings. The petrological features of these intrusions are similar from those in Western and Eastern

Himalaya. These intrusives are devoid of mafic xenoliths. Many of the leucogranites occur in calc silicate zones in the Higher Himalayan high grade terrains. Possibly these intrusions were emplaced after the MCT. According to Gansser (1993) there exists a gap of 500 m.y. between the neogene and anorogenic granites.

The largest uplift of the Himalya is exemplified by the Kailash. All major rivers such as Indus, Sutlej, Ganga, Karnali *etc*. oraginate from Kailash and substantiate vertical uplift. Its height is 6800 m and is also the Highest molasse mountain. Some of the Trans Himalayan plutons *e.g.* Quxu and Gandese were uplifted upto 3 km in late Oligcene / early Miocene (Copeland et al., 1987). Possibly the largest uplift was the Kailash while the Himalaya was still in its rudimentary stage dominated by the Trans Himalaya (Gansser, 1993).

Steck et al. (1993), Dezes (1999), Epard et al., (1995), Frank et al., (1973), Girard (2001), Robyr et al., (2002) and Schlup et al., (2003), emphasized about the over burden of the eroded rocks during Himalayan-tectonics based on many geological cross-sections of Zanskar, Yunam, Manali and Rampur areas of the western Himalaya (Fig. 4.4). It is an important issue and as per vertical scale it is estimated that about 30-40 km of the overburden of eroded rocks was removed. It is generally regarded that thrusting in the Himalaya took place about 20 m.y. ago and some dates are as young as 5 m.y. In such a short time how it would have been possible that 30-40 km thick overburden occurred and all this overburden was deposited in the Indian sea. Therefore these estimates appear to be more conjectural and exaggerated.

During the underthrusting the Himalayan rocks were detached from the Indian crust and accreted on the Asian backstop formed by Ladakh batholith migrating southwards through the Himalayan nappes upto the Lesser Himalaya and characterized by the Main Boundary Thrust (Fig. 4.5). This thrust of the Siwalik molasses is active and the seisimic activity has been reported at a depth of 10-15 km below the MCT by Avouac et al.(2001). The Himalayan orogeny is still continuing and a new thrust is in the making i.e. the Main Frontal Thrust which has been designated as the Active Himalayan Thrust by Steck (2003).

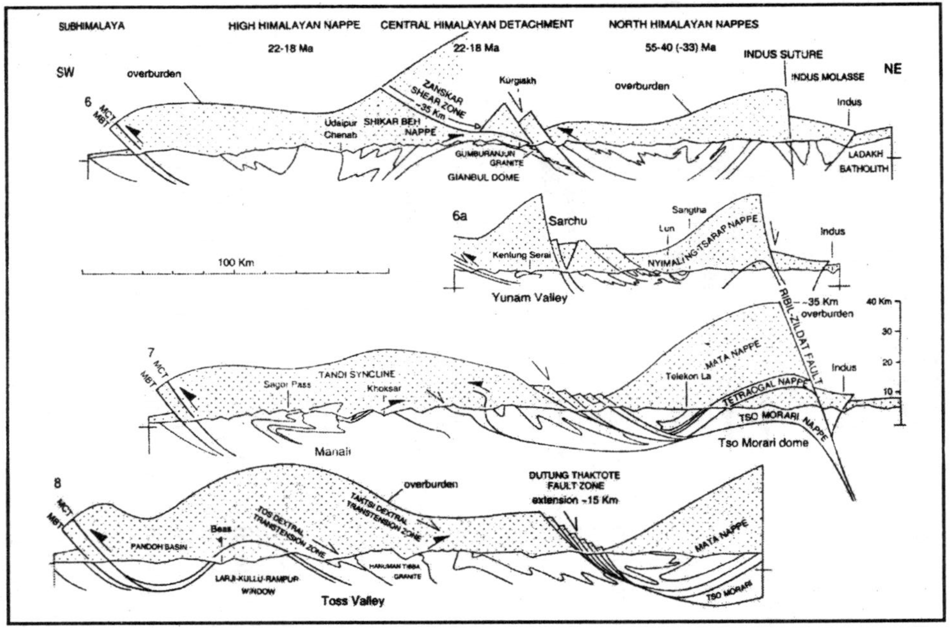

Figure 4.4: Estimates of the eroded overburden of the Himalayan accretionary wedge based on thermo-barometric data from Dezes (1999), Epard et al. (1995). Frank et al. (1973b). Girard (2001), Robyr et al. (2002), Schlup et al. (2003), and Steck et al. (1993). The numbers 6, 7and 8 correspond to the cross-sections in the Yunam Valley and the Tsara River near Lun and Sangtha. Note the strain partitioning of the Central Himalayan detachment along the strike of the Himalayan range. (from Steek, 2003)

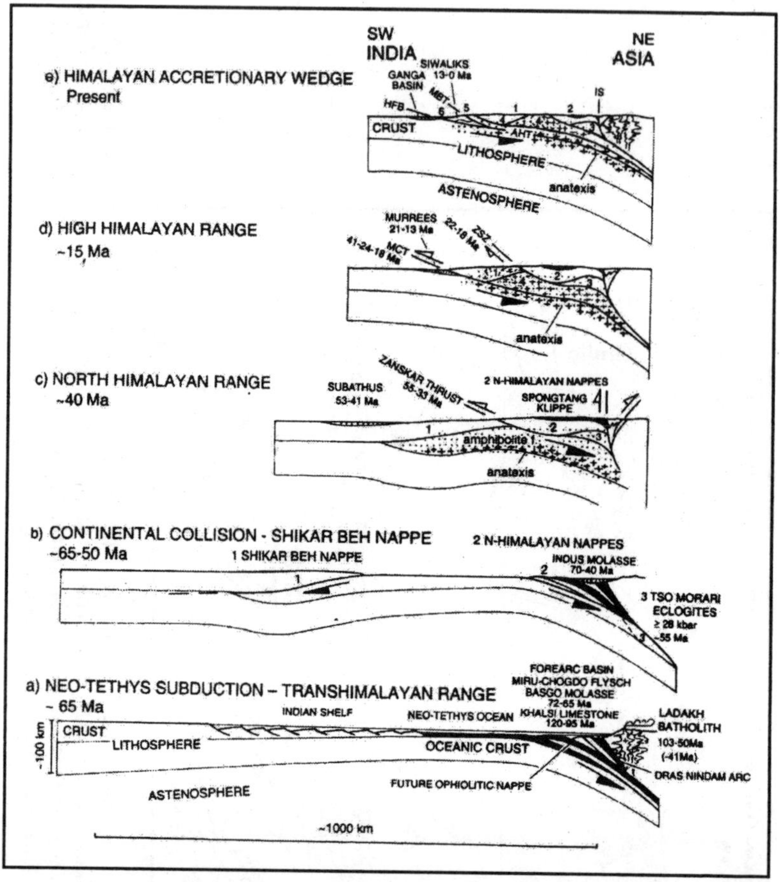

Figure 4.5 : Model for the kinematic evolution of the NW Himalaya, AHT = Active Himalayan thrust, HFB = Himalayan Frontal boundary, IS = Indus Suture, MBT = Main Boundary thrust. MCT = Main Central thrust, ZSZ = Zanskar shear zone (after Steck, 2003).

The main uplift of the Himalayan nappes took place between 15-40 m.y. and the rate was 2-5 mm per year. The Higher Himalayan uplift was initially slow but was increased during 11 and 21 m.y.

References

Auden, J.B. 1935; Traverses in the Himalaya, *Rec.Geol. Surv. India*, **66**(4), 123-167.

Avouac, J.P., Bollinger, Lave, J., Cattion, R. and Flouzat, M., 2001; Le Cycle Seismique en Himalaya, *C.R. Acad. Sci., Paris,* **333**, 53-529.

Bird, P., 1978. Initiation of intra continenetal subduction in the Himalaya, *Jour.Geop.Res.* **83**, 4975-4987.

Copeland, P., Harrison, T.M., Kidd, W.S.F., Ronghua, X. and Yuquan, Z. 1987; Rapid early Miocene acceleration of uplift in the Gangdese Belt, Xizang (southern Tibet), and its bearing on accommodation mechanism of India-Asia Collision. *Earth. Plan. Sci. Lett.,* **86**, 240-252.

Davidson, C., Grujic, D.E., Hollister, L.S., Schmid, S.M., 1997. Metamorphic reactions related to decompression and synkinematic intrusion of leucogranite, High Himalayan Crystallines, Bhutan. *Journal of Metamorphic Geology* **15**, 593-612.

Dezes, P., 1999, Tectonic and metamorphic evolution of the Central Himalayan domain in southeast Zanskar (Kashmir, India), *Mem. Geol.,* **32**, Lausanne.

Epard, J.L., Steck, A., Vannay, J.C. and Hunziker, J., 1995, Tertiary Himalayan Structures and Metamorphism in Kulu Valley (Mandi Khoksar transect of the Western Himalaya), Shikar Beh Nappe and Crystalline nappe, *Schw. Min. Pet. Mit.,* **75**, 59-84.

Fank, W., Hoinkes, G., Miller, C., Purtscheller, F., Richter, W. and Thoni, M., 1973, Relation between Metamorphism and orogeny in a typical section of the Indian Himalaya, *Tesch. Min. Pet. Mit.,* **20**, 303-332.

Gansser, A., 1964. *The Geology of the Himalayas,* Wiley Interscience.

Gansser, A., 1993. Facts and theories on the Himalaya, *Geol. Soc. Ind.,* **41**, 487-508.

Girard, M., 2001, Metamorphism and tectonics of the transition between nonmetamorphic tethyan Himalaya Sediments and North Himalayan Crystalline Zone & Rupshu area, Ladakh, NW India, *Mem. Geol. (Lausanne),* **35**.

Harris, N., Inger, S. and Massey, J., 1993, The role of fluids in the formation of High Himalayan leucogranites. In : *Himalayan Tectonics,* (Eds. Trealor, P.J. and Searle, M.P.), 391-400, Geol. Soc. Lond.

Harrison, T.M., Grove, M., Lovera, O.M. and Cattos, E.J., 1998, A model for the origin of Himalayan anatexis and inverted metamorphism, *Jour. Geoph. Res.,* **103**, 27017-27032.

Harrison, T.M., Grove, M., Lovera, O.M., Catlos, E.J. and Andrea, J.D., 1999. The origin of Himalayan Anatexis and Inverted Metamorphism, Models and constraints. *Jour. Asian Earth Sci.,* **17**, 755-772.

Heim, A. and Gansser, A., 1939, *Central Himalaya : Geological observations of the Swiss expedition,* 1936, Denk. Schw. natu. 73(1).

Hodges, K.V., Hubbard, M.S., Silverberg, D.S., 1988, Metamorphic Constraints on the thermal evolution of the central Himalayan orogen, *Phil. Trans. Roy.Soc. Lond., A-326,* 257-280.

Hubbard, M.S., 1996, Ductile Shear as a cause of inverted metamorphism : example from Nepal Himalaya, *Jour. Geol.,* **104**, 493-499.

Krummenacher, D., Basett, A.M., Kingery, F.A. and Layne, H.F., 1978, Petrology, metamorphism and K-Ar age determinations in Eastern Nepal, *Tectonic Geology of the Himalaya,* In : Saklani, P.S. Ed. 151-168, Today and Tomorrow's Publishers.

Le Fort, P., 1975, Himalaya, the collided range, Present knowledge of the continental Arc, *Am. Jour. Sci., 275-A,* 1-44.

De Loczy, L., 1907: Beobachtungen im ostlichen Himalaya. *Foldr. Kozlem.* **35/9**, 1-24.

Pecher, A., 1978: Deformations et metamorphisme associes a une de cisaillement. Example du grand chevauchement central himalayen (M.C.T.), transversalede Annapurnas et du Manaslu, Nepal. Thesis Doct. Univ. Grenoble.

Pecher, A. and Le Fort. P., 1986: The metamorphism in central Himalaya, its relation with the thrust tectonics., *Sci. de la Terre, Nancy, mem. 47,* 285-309.

Robyr, M., Vannay, J.C., Epward, J.H. and State, A., 2002, Thrusting, extension and doming during polyphase tectonometamorphic evolution of the High Himalaya Crystalline Zone, NW India, *Jour. Asian Earth Sci.,* **21**, 221-239.

Schulp, M., Carter, A., Cosca, M. and Stach, A., 2003, Exhumation history of eartern Ladakh revealed by 40 Ar / 39 Ar and fision track ages, the Indus river, Tso-Morari transect, NW Himalaya, *Jour. Geol. Soc. Lond.,* **160**, 1-15.

Steck, A., Spring, L., Vannay, J.C., Mason, H., Bucher, H., Stuiz, E., Merchant, R. and Tieche, J.C., 1993, Geological transect across the north western Himalaya, Eastern Ladakh and Asia, *Eclo. Geol. Helv.,* **86**, 219-263.

Steck, A., 2003, Geology of the NW Indian Himalaya, *Eclo. Geol. Helv.*, **96**(27), 147-196.

Valdiya, K.S., 1976, Himalayan transverse faults and folds and their parallellism with subsurface structures of North Indian plains. *Tectonophysics,* **32**, 353-386.

Valdiya, K.S., 1980, *Geology of Kumaun Lesser Himalaya.* Wadia Inst. of Himalayan Geol., Dehra-Dun.

Valdiya K.S., 1981, Tectonics of the Central Sector of the Himalaya, In : Zagros. Hindukush, Himalaya Gèodynamic Evolution (Gupta, H.K., ed.) *Am., Geoph. Unio, Ser.***3**, 87-110.

• □□□ •

5

Himalayan Seismicity

Plate tectonic evolution of the Himalaya is ascribed to collision of Indian – Eurasian plates (Lyon-Caen and Molnar, 1983, Molnar, 1986; Valdiya, 1984 a, b). During Cretaceous or Palaeocene time the Tethys was a part of Indian-Eurasian plate. The lithospheric thickening led to the formation of the Himalaya. According to Molnar (1986), the Higher Himalaya was uplifted with the formation of the MCT which shows a low angle dip at greater depths. Due to northward movement of the Peninsular plate, the Tethys – Sediments were also uplifted.

The tectonic transport of the Main Central Thrust (MCT) during 10 to 20 m.y. BP was rather slow. Subsequently the Main Boundary Thrust (MBT) became active and was subjected to uplift. The neotectonic activity of the Indian plate is still continuing and is evidenced by seismicity along the MCT and MBT.

Verma (2005) preferred to describe the seismicity of the arcuate Himalaya as Western and Eastern Himalayn arcs. The western arc is located between 25° to 36° N and 70° to 85° E. He (2005) also gave a map of the earthquakes with maguitude of 5.5 and above. According to him the Western Himalaya is seismically active in Kumaun, Garhwal, Panjab and Kashmir and it is of interest to note that majority of the earthquakes (magnitude 6 and above) occurred in the proximity of the MCT (Fig. 5.1).

NORTH-WESTERN HIMALAYAN SEISMICITY

This region is tectonically delineated by the MBT (= Riasi and Panjal Thrust) which gets terminated near Kashmir syntaxis but the MBT extends westwards in Pakistan and is known as the

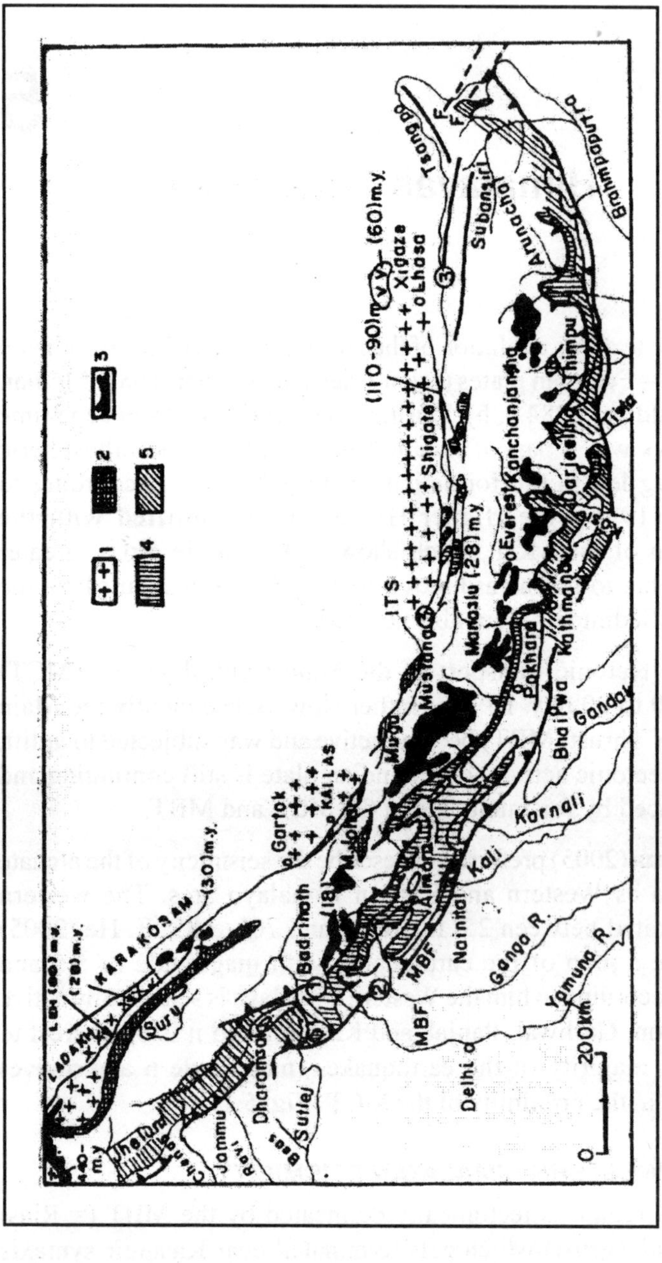

Figure 5.1: Simplified geologic map of Himalaya (after Gansser, 1974 and Valdiya, 1984b), showing the locations of major features such as the Tsangpo Suture (ITS) the Main Boundary Fault (MBF) the Main Central Thrust (MCT). 1. Granites, 2. Ophiolites, 3. Middle Tertiary granites, 4. Metamorphic nappes. 5. Epimetamorphic nappes.

Hazara thrust. Verma (1991, 2005) opined that the Hazara Thrust is seismically active. It is to state that the intensity of epicenters is more near Jammu, Simla, Chamba and Kinnaur areas. Ni and Barazangi (1985), Verma and Kumar (1987) and Chatterjee and Bhattacharya (1992) are of the view that seismicity was mainly controlled by thrust faulting and pull-a-part structures. (Fig. 5.2).

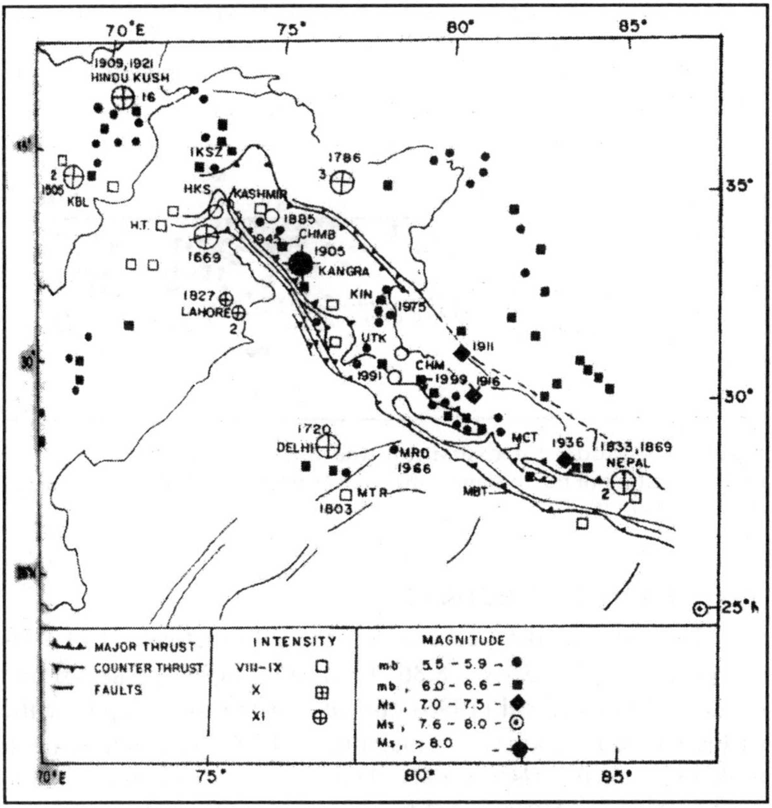

Figure 5.2: Epicentral map of western Himalaya showing the historical earthquakes of intensity VIII or magnitude 5.5 and above. The figure shows high seismicity along the Himalayan arc, the Hindukush region, the Gangetic plains as well as the Tibetan Plateau. Abbreviations : CHM - Chamoli, CHMB Chamba, HKS - Hazara Kashmir Syntaxis, HT-Hazara thrusts, IKSZ - Indus Kohistan Seismic Zone, KIN-Kinnaur, KBL-Kabul, KNG-Kangra, MBF-Main Boundary Fault, MCT-Main Central Thrust, MRD-Moradabad, MTR-Mathura UTK-Uttarkashi.

The Uttaranchal Himalaya is tectonically characterized by the MBT, North Almora Thrust (NAT) and MCT and was affected by several earthquakes (> 6 magnitude) in 1916, 1935 and 1945.

According to Khattri et al., (1989) and Khattri (1992) most of the earthquakes in Uttaranchal originated at a depth of less than 13 km and ascribed them mainly due to underthrusting of the Indian Shield in Lower and Higher Himalayan belts (Fig. 5.3).

The Uttarkashi and Chamoli earthquakes in recent years in Uttaranchal are of great geological significance.

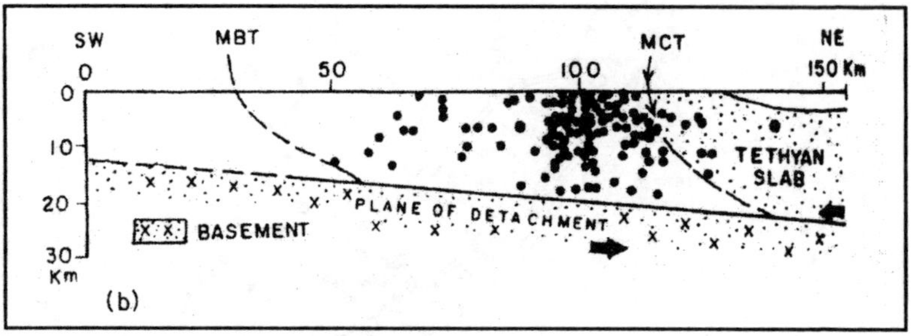

Figure 5.3: Model for explaining the microseismicity in terms of underthrusting of the Indian continental crust underneath the Himalaya (Khattri et al., 1989).

UTTARKASHI EARTHQUAKE

This earthquake occurred on the 20th October, 1991. Its epicenter was 30.75° N, 78.86° E. Rastogi (1995), Rastogi and Chadha (1995,) Kayal (1996) opined that the focal depth of the earthquake was 12 km with a magnitude of 6.6 on Richter Scale. It took place at 0.2 (H): 53(M). The maximum intensity of the earthquakes was VIII+ on MM scale. It was followed by as many as 40 after shocks. The peak ground accelerations were highest at Bhatwari and Uttarakashi (30% of g). The maximum duration of strong motion was more than 1% g for 40 seconds. The major and minor shocks were confined to the Main Central Thrust Zone indicating about reactivation of the thrust. Kayal et al., (1995) and Kayal (2001) gave cross-sections (Fig. 5.4 a, b) of the aftershocks and ascribed the Uttarakashi earthquake due to rupture of the Indian plate along a detachment surface (underthrusting) underneath the Lesser Himalaya. Seeber and Armbruster (1981) also hold a similar view. The Higher Himalaya underlies the Himalayan sedimentary wedge and due to rupture of Indian plate associated with northward movement gave rise to earthquakes along the Main Boundary and Main Central Thrusts.

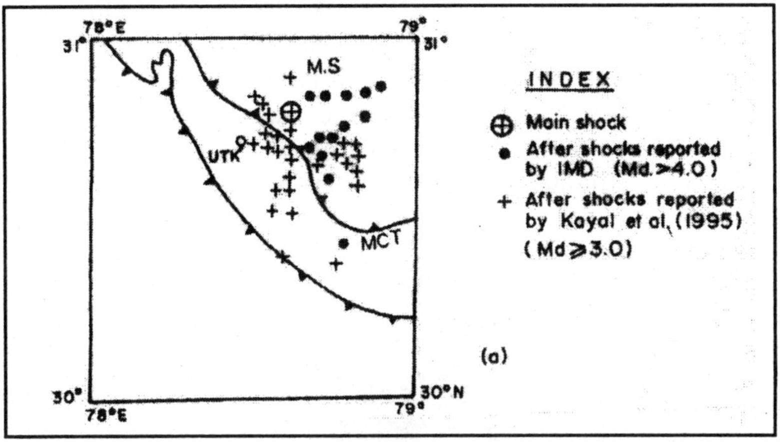

Figure 5.4 a: The figure shows the locations of the Uttarkashi area.

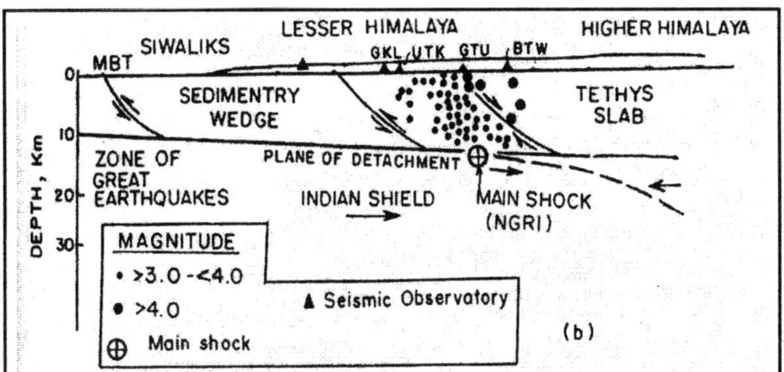

Figure 5.4b: The possible model of Himalaya in terms of underthrusting of the Indian lithosphere underneath the Lesser Himalaya (Kayal et al., 1995).

CHAMOLI EARTHQUAKE

Due to intraplate movement an earthquqake (Mag 6.8; IMD) occurred at Chamoli on 29th March, 1999. Its epicenter was between 30.4° N, 79.4° E with a focal depth of about 20 km. Rastogi (2000) opined that the rupture with intensity VI extended for about 250 x 150 km along NW-SE direction (Fig. 5.5 a, b). Thrust faulting along ENE-WSW was the main cause of this earthquake. The aftershocks of this event were studied by IMD, NGRI (Rastogi, 2000) and G.S.I., (Kayal et al., 2003). It may be remarked that the aftershocks were located towards north and south of the MCT. The model of Chamoli earthquake is similar to that of the Uttarkashi earthquake.

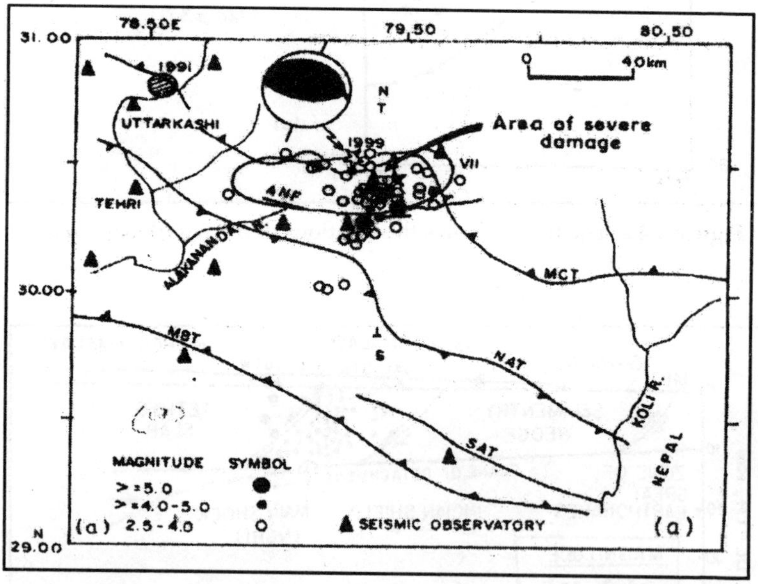

Figure 5.5a: Epicentral map of Chamoli earthquake (1999) and its after-shocks (after Kayal, 2001). ANF - Alakananda fault, MBT - Main Boundary Thrust, MCT - Main Central Thrust - North Almora Thrust, SAT-South Almora Thrust.

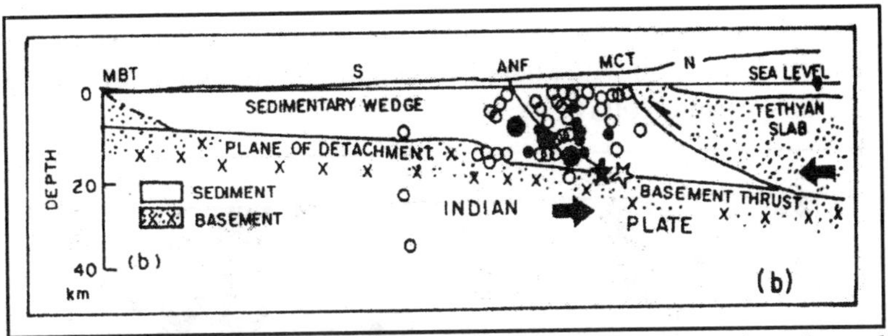

Figure 5.5b: Plate underthrusting model for the Chamoli earthquake and its after-shock, (after Kayal, 2001).

Nepal Himalayan Seismicity

Nepal is located between 80° to 88° E and 26.5° to 31° N. Chitraker and Pandey (1996) investigated the seismicity of Nepal. Many earthquakes of about 7 magnitude took place in Nepal in 1916, 1934 and 1936. Gautam et al. (2002 b) have given epicentral map (4 to 4.9 magnitude) of this region based on IMD data (Fig. 5.6). It is clear from the map that the sections along A-A', B-B' and C-C' covering the MBT to Tibetan plateau are seismically active. Bilham et al., (1997) remarked that the uplift of in the Higher Himalaya is 5.7 mm per year and in the Kathmandu valley region the horizontal movement is about 42 mm per year. The rate of uplift of the Everest is 3-5 mm per year. The thrust systems accommodate nearly 18.7 mm per year due to convergence of the Indian and Eurasian plates.

EASTERN HIMALAYAN SEISMITY

The Sikkim, Bhutan and Shillong plateau, Assam, Mikir hills and their adjoining parts are seismically active. Gautam et al. (2002) have shown the epicentral locations (4 to 4.9 magnitude) of the mentioned areas (Fig. 5.6). Verma and Kumar (1987) are of the view that in several regions the focal depth along the faults reached upto 50 to 80 km.

Like western Himalaya the MBT and MCT Zones are seismically active in Eastern Himalaya (Fig. 5.6). Close to MCT many earthquakes of, magnitude 5 and above have occurred. Earthquakes of 6.5 and 7.1 magnitude took place in Mikir Hills in

1941. Shillong plateau was affected by a major earthquake of 7.1 magnitude in 1930 while near Dauki fault the shock was of 7.1 mg in 1923 (Verma, 2005).

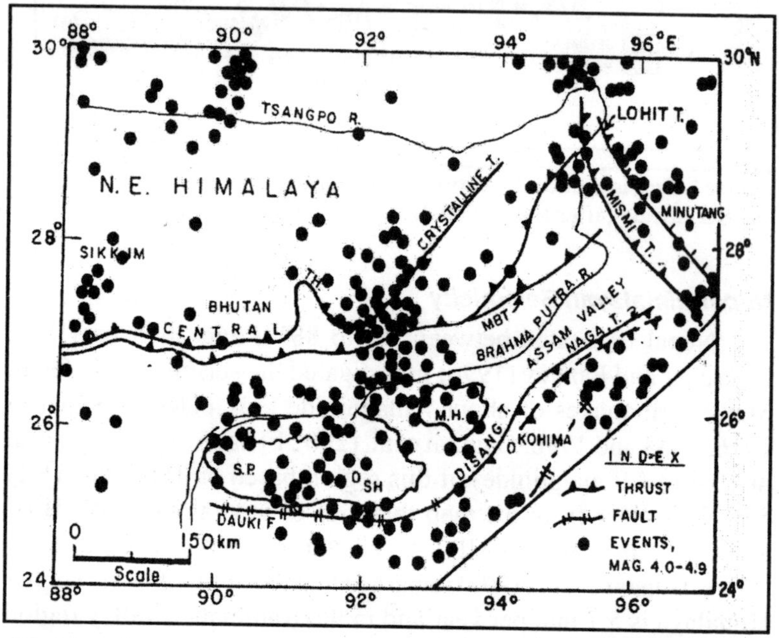

Figure 5.6: Seismicity of north-east India and adjoining regions showing the locations of events of mag. 4.0-4.9 (period 1963-2000) (after Gautam et al. 2002). The figure shows high seismicity over the Shillong Plateau, Eastern Himalaya and the Mishmi Thrust-region. Only the upper Assam Valley is relatively inactive. Abbreviations: Koh-Kohima, M.H. - Mikir Hills, SH-Sillong, S.P. - Shillong Plateau.

Mishmi and Lohit thrusts trending NW-SE are present in the syntaxial region. Assam earthquake of 8-7 magnitude took place towards east of the Lohit thrust in 1950. Gautam et al., (2004) described about the earthquakes of 4 to 4.9 magnitude in Shillong plateau and Mishmi thrust region (Fig. 5.6).

Areas of rupture along the Himalaya due to earthquakes have been described by Verma (2005) who also modified the map of Seeber and Armbruster (1981). There exists a major seismic gap between the Kangra (1905) and Bihar (1934) earthquakes. The earthquakes have taken place in Uttaranchal in 1916, 1934 and 1936. The potential area of rupture with a magnitude of 8 or above does exist between the Kangra and Bihar earthquakes and also in Kashmir Himalaya.

MODELS OF HIMALAYAN OROGENY

Verma (2005) gave subduction Zone model of the Himalaya (modified after Seeber and Armbruster, 1981) which depicts the subduction of the Indian plate underneath the Lesser and the Higher Himalaya upto the Tibetan plateau (Fig. 5.7).

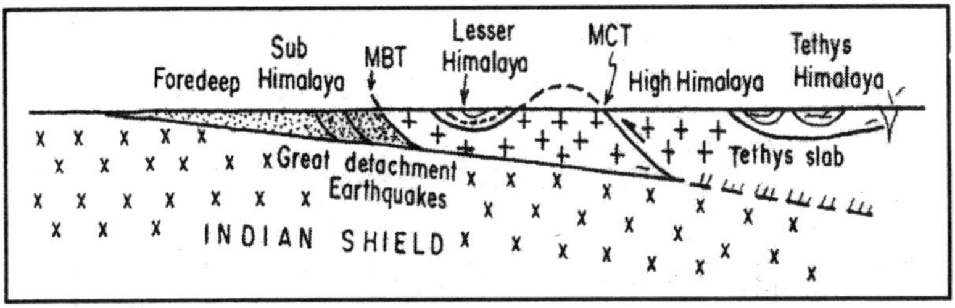

Figure 5.7: Schematic model of continental subduction along the Himalayan arc (modified after Seeber and Ambruster, 1981). The model can explain satisfactorily the present seismicity close to the MBF or the MCT.

It has been established that major seismic activity is taking place in the areas close to the MBT or the MCT. The modified model after Seeber and Arumbruster (1981) explains the recent earthquakes of Uttarkashi and Chamoli but it can not explain satisfactorily the seismicity of the Shillong plateau and the Mishmi thrust zone.

Acharyya (2005) discussed about the Indian-Asian plates interaction and proposed the evolution of the Himalaya by a two stage process; (a) fragmentation of the Pan Indian continent during late Mesozoic with creation of narrow peri-Indian ocean flanked by epicontinental fragments and (b) closure of oceans with collisonal amalgamation of rock-fragments. The collision also occurred in two stages. The initial collision of Kohistan-arc with the Tibetan margin (Karakorum-Gangdise) in Palaeocene led to accretion of ophiolites. During late Eocene the continent continent collision accreted Indus Tsangpo Ophiolites and its cover sediments were thrust over the folded flysch and shelf sediments of the Tethyan Himalaya. Due to jamming of the Indus-Tsangpo suture zone and post-collisional northern movement of Indian plate several thrusts were developed in the Himalaya. Acording to him the southern margin of the Himalaya was characterized by the earlier formed MBT than the MCT. The Eocene-early neogene foreland and hinterland sediments and nappes were involved in late stage thrusting, duplex and schuppen structures. The model for the Himalayan evolution by Acharyya (2005) is given in (Fig. 5.8).

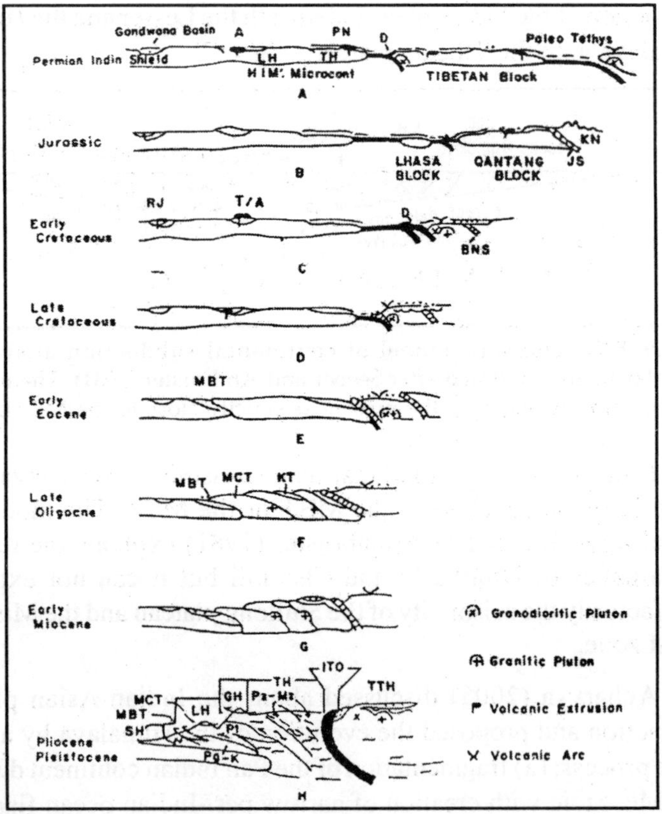

Figure 5.8: Tectonic evolution of the Himalaya and adjacent region. Oceanic crust shown by solid black line and by mossaic in inset, island arc shown by a hump Inset and Plio - Pleistocene cartoons show enlarged views. Inset shows folded oceanic crust with blocked subduction. Subduction dismembers the oceanic crust and causes its emplacement and is also affected by later plutons.Additional abbreviation. JS - Jinsha Jiang Suture, KB (inset) - Karakorum block,KN - Kunlun Range, LH - Lesser Himalaya, RJ - Rajmahal Volcanics, SH - Sub Himalaya, TH - Tethyan Himalaya, T-Volcanics of Tansen area, TTH - Tibetan Trans-Himalaya. Stratigraphic symbols used. Pt - Proterozoic, Pz - Palaeozoic, Mz - Mesozoic, Pg-K - Palaeogene - Cretaceous, N - Neogene. (After Acharyya, 2005)

According Dubey and Bhat (2005) the Indus Suture, the MCT and the MBT are manifestations of earlier basement faults formed during the birth of the Tethys which were rejuvenated during the compressional forces of the Himalayan orogeny.

Sharma (2005) proposed the model of Subduction zone mainly based on space and time distribution of various igneous rocks from Kohistan, Ladakh and Lhasa areas of the Indus collision zone. The Indus suture was generated earlier than 120 Ma in Lhasa (Pudsey et al., 1984) and was older in Kohistan-Ladakh sector. The presence of 156 Ma Dras rocks in Ladakh and volcanics (135 Ma)in Lhasa suggest that the older oceanic crust to a great extent was subducted and consumed to give rise a widespread magmatism (Fig. 5.9).

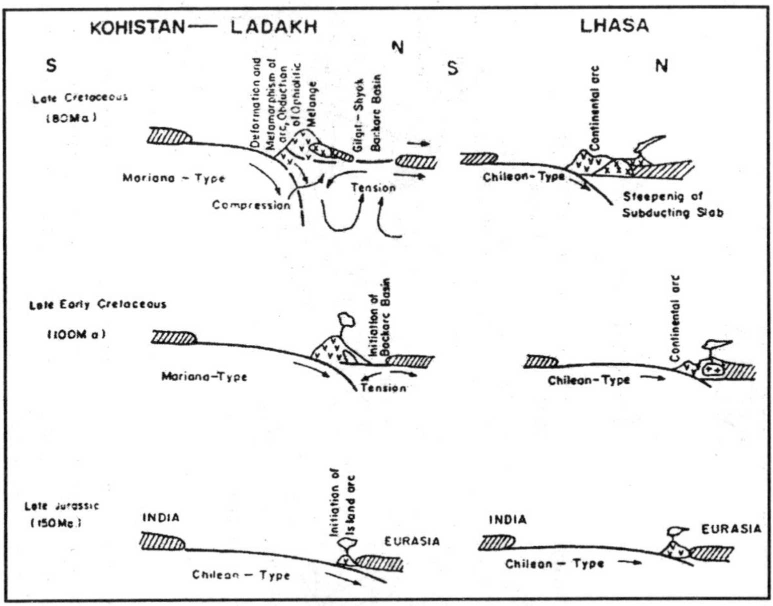

Figure 5.9: Diagrams showing changing mode of subduction and development of island arc, continental arc and back arc basin. (After Sharma, 2005).

Steck (2003) described the geology of the Northwestern Indian Himalaya and gave a model of its main tectonic units. (Fig. 5.10). He elucidated that the Himalayan evolution started at (1) 65 m.y. (2) 65.50 m.y. (3) 40 m.y. (4) 15 m.y.

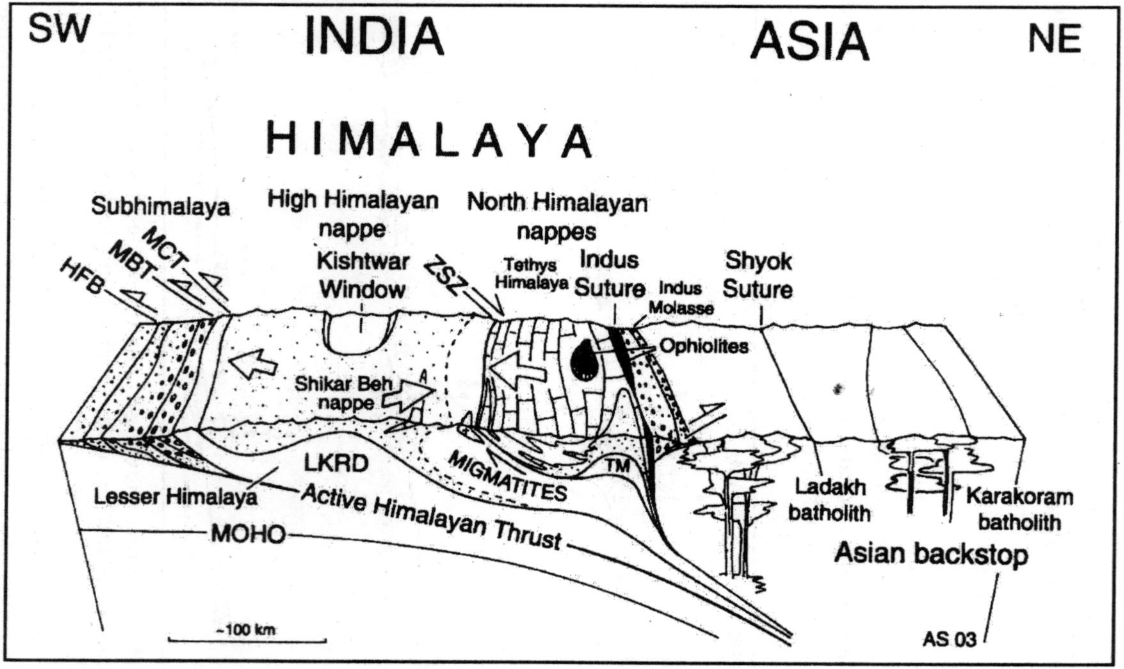

Figure 5.10: The main tectonic units of the NW Indian Himalaya HFB = Himalayan Frontal Boundary, LKRD = Larji-Kullu-Rampur dome MBT = Main Boundary Thrust, MCT = Main Central Thrust, ZSZ = Zanskar Shear Zone (After Steck, 2003)

Catlos et al., (2004) investigated the MCT in Sikkim region of eastern Himalya and regarded the thrusts to be out of sequence (Fig. 5.11). The intrusive tourmaline granites of the Himalaya (14.8 ± 0.3 Ma and 2.0 Ma) are products of melt formation within the Himalayan orogenic belts. Leucogranites are exposed near Gangotri, Manaslu, Sikkim *etc.*

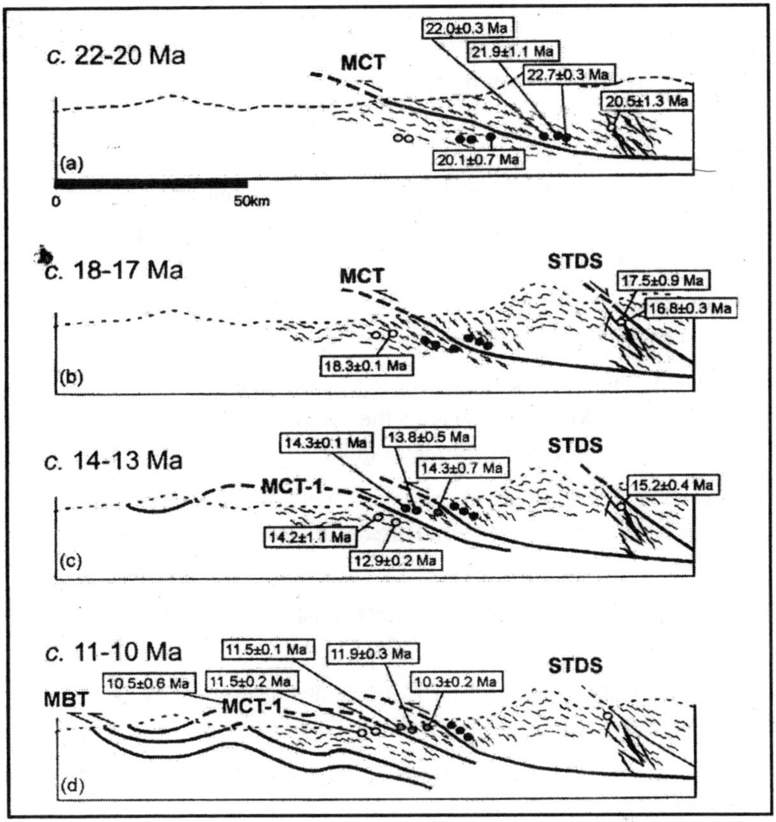

Figure 5.11: Schematic interpretation of the age results and P-T constraints from Sikkim rocks. Greater Himalayan Crystallines rocks (samples NLG 963, LCG 542 and LCG 541) are indicated by black circles, and the Lesser Himalaya samples (KBP 1062A and KBP 1062 C) are white. The High Himalayan leucogranite MK 51K is indicated in white as well. (After Catlos et al., 2004).

Th-Pb monazite ages indicate that MCT in Sikkim was active at about 22, 14-15 and 12-10 m.y. These monazite ages decrease towards lower structural levels in the lesser Himalaya. The monazite age varies as 11.5 ± 0.2 to 12.2 ± 0.1 Ma. This age from sheared low grade metamorphics of the MCT ranges from 13.8 ± 0.5 11.9 ± 0.3 Ma.

Saklani (1993) based on his investigations in Garhwal Himalaya established that the MCT zone is characterized by imbrications or schuppen structures and the varying ages of the crystallization are due to imbrications which pile over one another.

References

Acharyya, S.K., 2005. The Himalaya and the Indian - Asian Plates interactions, In : Saklani P.S. (ed.), *Himalaya (Geological Aspects)* **2**, 275-293.

Bilham, R., Larsen, K. and Freymueller, 9., 1997.Indo-AsianConverquence rates in Nepal Himalaya, *Nature,* **386**, 61-66.

Catlos, E.J., Dubey, C.S., Harrison, T.M. and Edwards, M.A. 2004, Late Miocene Movement within the Himalayan Main Central Thrust Shear Zone, Sikkim,Northeast India, *Jour. Met. Geol.,* **22**, 207-226.

Chatterjee, S.N. and Bhattacharya, S.N., 1992, Micro-earthquake surveys and seismicity in Western Himalaya, *Geol. Soc. Ind. Mem.,* **23**, 23-44.

Chitraker, G.R. and Pandey, M.R., 1996. Historical earthquakes of Nepal, *Bull. Geol. Soc. Nepal,* 6, 7, **8**.

Dubey, A.K. and Bhat, M.I., 2005. Listric faults and the Himalayan evolution, In : Saklani, P.S. (Ed.), *Himalaya (Geological Aspects),* **2**, 159-184.

Gansser, A., 1974.Mesozoic Cenozoic Orogenic Belts, *Geol. Soc. Lond. Spe. Pub.,* **4**, 267-278.

Gautam, J.L., Verma, R.K. and Gupta, H.V., 2002, Analysis of Seismicity of NW Himalaya and Seismic risk in th region, special volume, *Ind. Geol. Assoc.,* **35** (27), 17-26. (Eds. Gupta, L.N. and Gill, G.S.).

Kayal, J.R., 1996, Earthquake source parameters in north east India : a review, *Jour. Him. Geol.,* **17**, 53-60.

Kayal, J.R., 2001, Micro earthquake activity in some parts of the Himalaya and the tectonic model, *Tectonophysics*, **339**, 331-351.

Kayal, J.R., Ghosh, B., Chakraborty, P. and De, R., 1995, Aftershock study of Uttarkashi earthquake of Oct. 20, 1991, *Geol. Soc. Ind. Mem.*, **30**, 25-41.

Kayal, J.R., De, R., Chakraborty, R. and Karunakar, G., 2003, Aftershock Sequence of Chamoli Earthquake of March 29, 1999 in Garhwal Himalaya, *Bull. Seis. Soc. Am.,* **93**, 109-113.

Khattri, K.N., 1992, Local Seismic investigations in Garhwal Himalaya, *Geol. Soc. Ind. Mem.,* **23**, 45-66.

Khattri, K.N., Chander, R., Gaur, V.K., Sarkar, I. and Kumar, S., 1989, New Seismological results on the tectonics of Garhwal Himalaya, *Earth. Planet. Sci. Lett.,* **98**, 91-100.

Lyon-Caen, H. and Molnar, P., 1983, Constraints on the Structure of the Himalaya from analysis of gravity anomalies and flexural model for lithosphere, *Jour. Geoph. Res.* **88**, 8171-8191.

Molnar, P., 1986, The structure of the Mountain ranges, *Sci. Am.,* **254**, 70-79.

Ni J. and Barazangi, M., 1984, Seismotectonics of the Himalayan Collision Zone : geometry of the underthrusting Indian Plate beneath the Himalaya, *Jour. Geoph. Res.,* **89**, 1143-1163.

Pudsey, C.G. Coward, M.P., Luff, I.W., Shackleton, R.M., Windley, B.F. and Jan, M.Q., 1984, Collision zone between the Kohistan Arc and the Asian Plate in NW Pakistan, *Trans. Roy. Soc. Edinburgh*, **72**, 463-479.

Rastogi, B.K., 1995, Seismological Studies of Uttarkashi earthquake of Oct. 20, 1991, *Geol. Soc. Ind. Mem.,* **30**, 43-50.

Rastogi, B.K. and Chadha, R.K., 1995, Intensity and isoseismals of Uttarkashi earthquake of Oct. 20, 1991, *Geol. Soc. Ind. Mem.,* **30**, 19-24.

Rastogi, B.K., 2000, Chamoli-earthquake of magnitude 6-6 March, 29, 1999, *Jour. Geol. Soc. Ind.,* **55**, 505-514.

Saklani, P.S., 1993. *Geology of Lower Himalaya (Garhwal).* International Books.

Seeber, L. and Armbruster, J.R., 1981, Great detachment earthquakes along the Himalayan arc and longterm forecasting, In : Earthquake prediction, an International Review, *Am. Geoph. Union Pub., Ser.*, **4**, 259-275.

Sharma, K.K., 1975, Magmatism and Crustal growth in the Himalaya, In : Saklani, P.S. (ed.), *Himalaya (Geological Aspects),* **1**, 63-82, Satish Ser. Pub. House, Delhi.

Steck A., 2003. Geology of NW Indian Himalaya. *Scl. Geol. Helv.,* **96** (2), 147-196.

Valdiya, K.S., 1984 a, *Aspects of Tectonics,* Tata Mc Graw Hill Pub.

Valdiya, K.S., 1984 b, Tectonics of folded fringe of the Indian Platform, *Proc. 27th Int. Geol. Cong, Moscow*, **5**, 110-135.

Verma, R.K. and Kumar, G.V.R.K., 1987, Seismicity and nature of plate movement along the Himalayan Arc, NE India and Arakan Yoma : A Review, *Tectonophysics*, **134**, 153-175.

Verma, R.K., 1991, *Geodynamics of Indian Peninsula and the Indian Plate Margin,* Oxford IBH Pub., 228-301.

Verma, R.K., 2005, Seismicity of the Himalaya, NE India and Burma, In : Saklani, P.S. (ed.), *Himalaya (Geological Aspects),* **1**, 253-273.

• □□□ •

6

Tectonic Significance of Minor Structures of Garhwal

Chronologic and kinematic integration of the Mesoscopic structures along the Bhatwari - Tehri - Narendranagar in Garhwal Himalaya was carried out by Schwan and Saklani (1991; Fig. 6.1).

They opined that the orogenic configuration of the prolific Himalayan Mountains can be properly understood not only by elucidation of the chronologic questions but also by the study of the kinematic analysis such as (i) deviation of one structural generation from its normal strike direction, (ii) convergence of fabrics e.g., lineations of various generations of structures and (iii) mode of formation of folds and/or fold hinges and lineations in *a* (= x)-direction deviating from the standard genesis of fold and axis parallel to *b* (= y)-direction. These and other special factors for the interpretation of structures concerning the Himalaya are :

a) Older fabrics such as F_1 and F_2 are sometimes not developed in some areas due to folding-hostility of rock materials or feeble stresses and subsequent overprinting or obliteration of F_1.

b) On the other hand, strongly developed early folds F_1 with L_1 in competent materials, like quartzite, hardly allow or even prevent a later folding and/or cross-folding (*e.g.* F_1/F_3). Likewise, F_4 / S_4 fabrics are rarely developed or absent in gneissic or other competent rocks. Within incompetent schistose layers the younger structures (F_3/L_3 and F_4/L_4) are formed due to strong overprinting or obliteration of older structures.

c) As mentioned under a) and b) about the linkage of some structural types with rock properties it may be remarked that

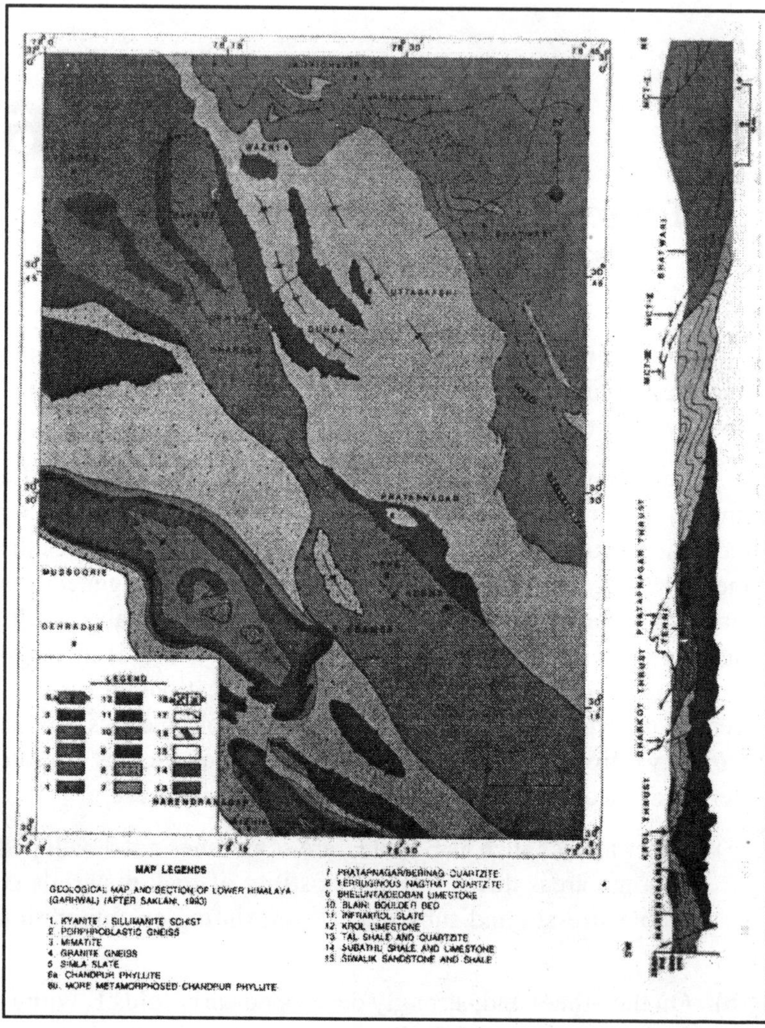

Figure 6.1:

B_5 structures observed in the entire Himalaya are not always developed in a single area. This fact is important for understanding the various of deformation phases in the Himalayan areas.

d) Primary deviations of directions of F_1 and F_2 fold axes from their normal NNE trend were noticed at some places. In fact, N to NNW and ENE trending fabrics occur in Extra-Peninsular and Peninsular regions.

e) Secondary deviations of the various structures can be observed in recumbent and plunging folds, which may be ascribed to hinge rotations. In particular, recumbent F_3 folds show axial scattering from the Himalayan longitudinal direction to transversal direction.

f) The folding along *a* (= x)-direction *i.e.*, the nappe transport to SSW, is kinematically problematic. Fold fabrics in *a*-direction are generally small and irregular and deviate from well developed normal folds in *b* (= y)-direction associated with cleavage. Fabrics along *'a'* were described from the Caledonides and Alps as sheath folds.

g) The normal *b*-lineations (streaming,) are parallel to the fold axis and *a*-lineations (striae on slickensides) represent the tectonic transport. The former ones are penetrative. The latter ones, in general, are found in rocks confined to thrust zones within the nappes.

h) The original directions of fold axes belonging to various generations within the overthrust masses can be completely changed by young sliding nappe tectonics and associated rotation. Schwan and Ray (1987) reported such an example from the Tista area (Sikkim) whereby the old cross folds F_1 + F_2 show rotation at about 90° into the longitudinal direction (NW) and the young longitudinal folds F_3 + F_4 occur in transversal direction (NE).

Theoretically, in the older rocks of a mountain range, all types of shortening structures (B_1 to B_5) can be developed. However, in younger rocks only the younger structures (B_3 to B_5) are present. It implies that earlier originated structures (B_1 + B_2) are confined only in older rocks. Symbols used for planar and linear structures are given in Fig. 6.1a.

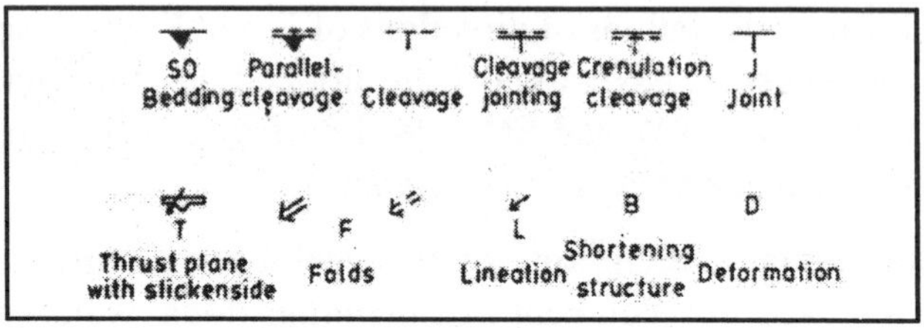

Figure 6.1a: Symbols of tectonic characteristics used in text and structural sketches.

Near Bhatwari, the cross-folds developed within augen-gneisses, garnetiferous schists and amphibolites contain F_1 and L_1 indicated by the positions of fold axes with lineations trending 27/36NE, 26/37NE, 15/37NNE and 10/31NNE (Fig. 6.2, on the right, 6.3 A, B, C; 4, 5, 6).

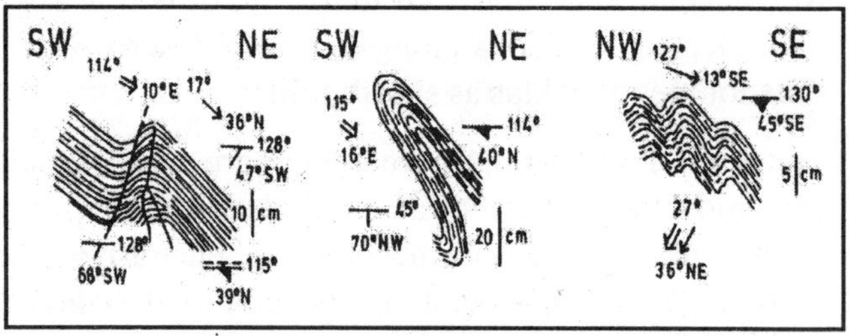

Figure 6.2: Inclined and overturned longitudinal folds as well as cross folds in migmatised augen gneisses and biotite schists; 200 m north of Bhukki, ca. 8, 5 km towards NE of Bhatwari.

With some NE lineations, besides L_1, also *a*-lineations due to nappe sliding (D_3) are present in the Chamba area (Dalhousie) in Himachal Pradesh (Schwan, Sinha and Bagati, 1985). The folded boudins in the quartzites and gneisses of the surrounding of Bhatwari show typical NE plunge (Fig. 6.3 A, B, C).

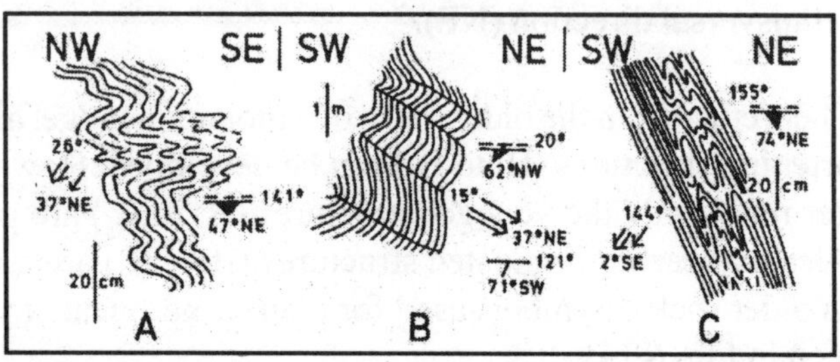

Figure 6.3: (A + B). NE striking folds, and (C): NW striking folds in grey garnetiferous schists; 7 km towards NE of Bhatwari.

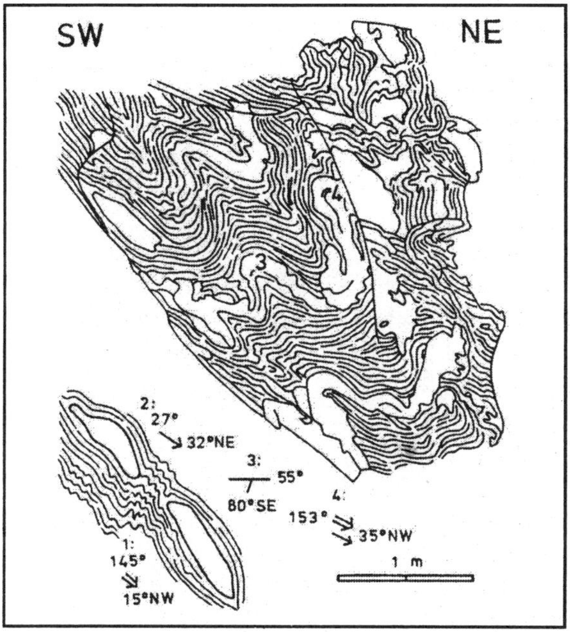

Figure 6.4: Overturned SW - vergent folds and boudinage structures with strong NE trending lineation are developed in amphibolite schists; 4.5 km NW of Bhatwari.

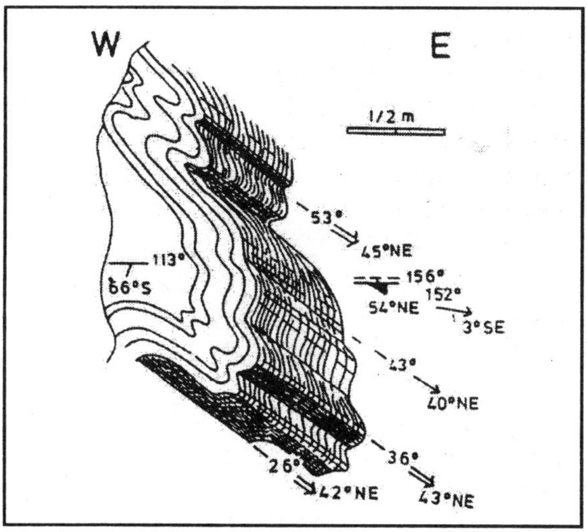

Figure 6.5: NE trending folds axes in amphibolite schists; 4 km towards NE of Bhatwari.

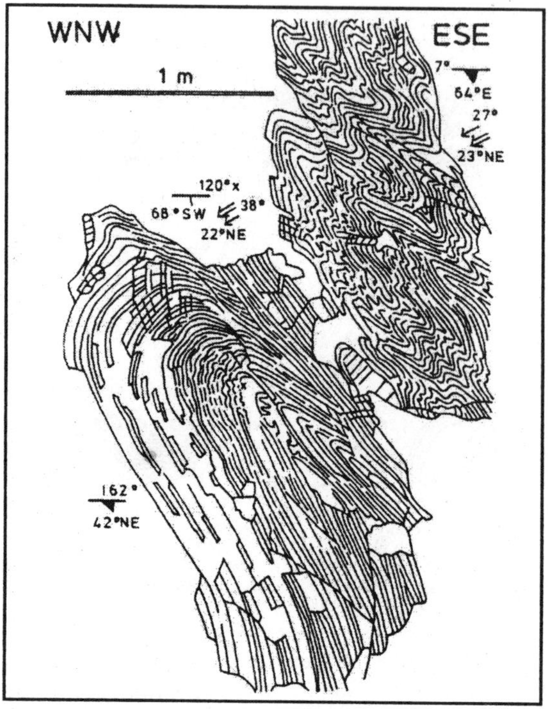

Figure 6.6: NE trending fold axes in amphibolite schists; 3.3 km towards NE of Bhatwari.

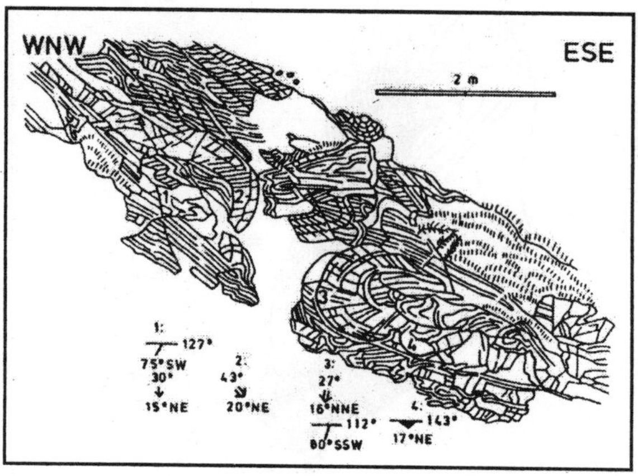

Figure 6.7: Quartzitic boudins in gneiss and amphibolite. Fold hinges and streaming lineations show NE trend and plunge. F_1 cross folds or rotated F_3; 1 km towards S of Bhatwari.

The main tectonic boundary between Great and Lesser Himalaya, marked by the MCT III corresponds as one of the most important movement planes in the Himalaya . Minor F_3 structures being very typical for the movement pattern of nappe tectonics are prominently observed in the flat recumbent folds of the Bhelunta Limestone in the Dunda Window, showing the characteristic Himalayan trend. Recumbent folds occur in the associated Bhainga Slates with thin quartzite layers. However, they show partly a deviating and across the strike position. These folds were rotated due to their flat axial planes and thus represent scattered F_3 folds.

In the area of the Chail Nappe 3 around Tehri lying between the Dharkot Thrust near Dharasu and the Pratapnagar Thrust near Chamba, uniform flat beds are present. However, like an exception a large mesoscopic structure with overturned limbs and with normal NW axial course represents a distinct example for F_3, and further few smaller structures trending towards N near Chamba are present and represent F_3 folds.

The chevron folds with crenulation cleavage in Simla Slates and Chandpur Phyllites dominating in the zone of parautochthony are typical F_4 / S_4 fabrics (Fig. 6.8, 6.9, 6.10). Perhaps the D_4 caused the upfolding of the Simla Slates and Subathu Shales in the Phart Window. The F_4 fold axial planes in Chandpur Phyllite near Hindolakhal are abnormally flat and F_4 developed due to younger tilting.

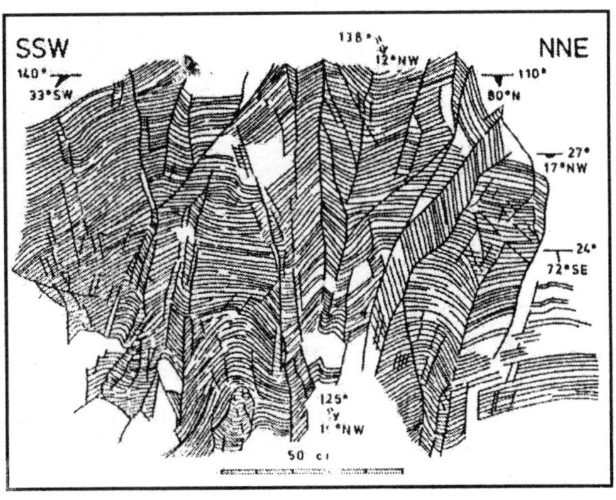

Figure 6.8: Chevron folds F_4 and crenulation cleavage S_4 with vergence towards SSW or NNE; Simla Slates along the road slope at the exit of NE of Jajal.

The upright and inclined folds in quartzitic bands of Infrakrol Slates near Narendranagar are strongly marked by perfectly developed F_3 folds (Fig. 6.1 A, B). The kink bands in Infrakrol Slates are the F_4 fabrics (Fig. 6.12).

The impact of D_5 deformation is represented by anticlinal breaking-up of younger transversal swelling of (Fig. 6.13).

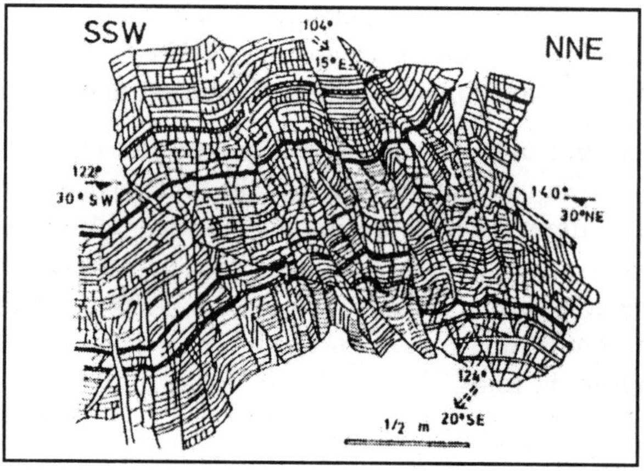

Figure 6.9: Kink band folds with spaced cleavage (F_4 + S_4); Chandpur Phyllite, northern exit of village Agra Khal (14 km north of Narendranagar).

Figure 6.10: South-vergent kink band folds (F_4) with peculiar flat position of axial planes in Chandpur Phyllite; Hindolakhal, 5 km north of Narendranagar.

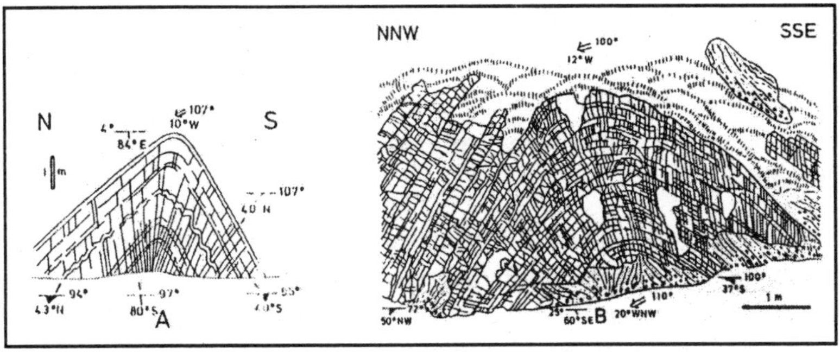

Figure 6.11(A) : Inclined and weakly south-vergent anticline with a distinct cleavage fan (F_3 + S_3 types). **(B)** : Inclined north-vergent anticline with fracture cleavage in the fan(F_3 + S_3 types). Quartzite beds of Infra-Krol; road slope at the southern exit of Narendranagar.

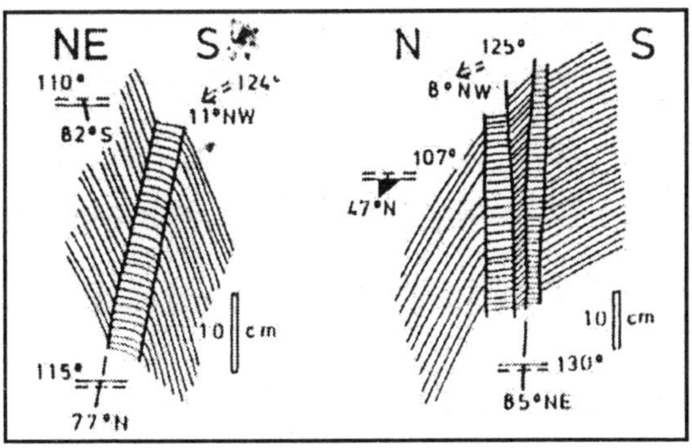

Figure 6.12: Kink band structures (F_4) in slate-quartzitic beds of Infra-Krol; 6 and 7 km south of Narendranagar.

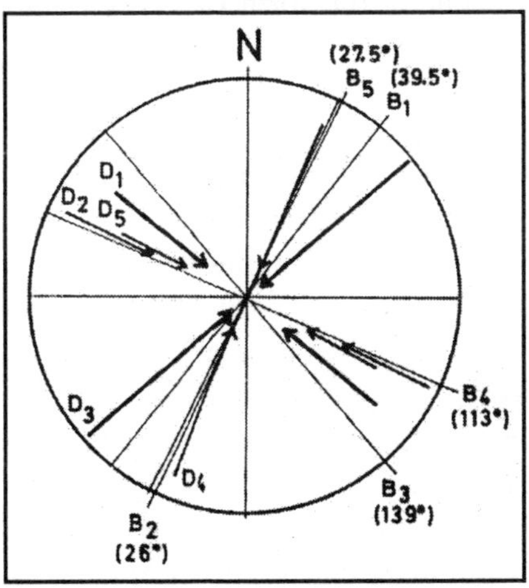

Figure 6.13: Collective diagram of the age sequence of the B_1 to B_5 shortening structure generations and / or the D_1 to D_5 compressional deformation vectors (Schwan 1981).

Kinematics of the Structures

Along the traverse between Bhatwari and Narendranagar several mesoscopic structural types are present which correspond chronologically and kinematically to the major tectonics of the Lesser Himalaya.

Visible mesoscopic fabrics in megascopic structures give evidence about the formal character and kinematics of the major structures (Schwan 1964, 1973, 1981). This function of small-scale structures for understanding the large-scale tectonics (e. g. nappe-tectonics) and its movement history was also stressed by Ramsay (1981) who inferred the strain and other structural features of the Helvetic Nappes in the Alps.

It is thus clear that in the Lesser Himalaya two main types of key-structures are present: (1) nappe thrusts and major overturned to plunging folds (D_3)corresponding to mesoscopic T_3 and F_3 structures, e. g. the flat movement planes and recumbent folds in the windows of Uttarkashi and Dunda and (2) the parautochthonous inclined major fold formation of the Krol Syncline developed during D_4 shows similarity with small-scale inclined structures of F_4 and S_4 (Fig. 6.8, 6.9).

NAPPES

The crystallines and metamorphics exposed between Bhatwari and Sainj (near Uttarkashi) occur in the form of Jutogh Nappe (6.1).

Jutogh Nappe

It consists of medium grade gneisses, migmatites and schists. The garnetiferous mica, schists and gneissic migmatites are exposed at Bhatwari. The grey to brown gneissic migmatites are mainly composed of biotite, plagioclase, and iron oxides. The gneissic migmatites are tightly folded and deformed in which palaeosomes of amphibolite and granite gneiss occur as undigested fragments. The amphibolites at times measure for about 2 metres in thickness. The biotite rich gneissic migmatites are exposed at Bhatwari and are frequently characterized by calc silicate bands. Quartz-chlorite schist bands occur at about 1 km towards southwest at Bhatwari.

The amphibolitic migmatites are exposed at about 250 metres southwards of Bhatwari and dip 20° - 30° SW and NE trending synform is bordered by the Main Central Thrust II, *i.e.* the Jutogh Thrust.

A tongue of biotite gneiss is exposed at Malla which is about 6 km southwards of Bhatwari along the motor road. It is associated with a metabasic sill and is traversed by a transverse fault. The main host rock of the migmatites is the biotite-gneiss which is exposed at Lata (about 2.5 km -southwest of Malla). Biotite gneiss is medium to coarse grained, greyish green and brown. It is mainly lensoidal and is exposed at Sainj which is at the contact of Main Central Thrust III (Budhakedar Thrust). The biotite gneiss is much folded around Sainj.

Budhakedar Nappe (Fig. 6.1)

The Budhakedar Thrust (MCT III) is well exposed at Sainj, near Uttarkashi which brings the low grade chlorite schists and granite gneisses to pile over the massive white Pratapnagar (Berinag) Quartzites. These crystallines and metamorphics together constitute the Budhakedar Nappe.

Pratapnagar Nappe (Fig. 6.1)

It is mainly constituted by the massive white quartzites characterized by intercalations of sericite -quartz schists. Metabasics are also associated in it. The quartzites are fine to coarse grained and are greyish green. Current bedding, graded bedding and mullions are observed in the quartzites. The mesoscopic folds of overturned type with associated thrust-faults are well developed in the quartzites near Asena and point towards their imbrication during the Pratapnagar thrusting.

Dharkot Nappe (Fig. 6.1)

The greyish green chlorite-sericite-quartz phyllites constitute the nappe and are well exposed at Dunda, Dharasu and Tehri. The folded phyllites of the Dharkot Nappe overlie the parautochthonous group exposed around Dunda and north of Tehri. There are 4 generations of folding present in the phyllites. The F_1 folds are asymmetrical to isoclinal and plunge 10-30° in SW-NE. The F_2 folds are open to recumbent and plunge 15-50° in WSW-ENE. The F_3 folds (10-40° NW-SE or WNW-ESE) show normally well developed recumbent or overturned types with penetrative flat cleavage. The F_4 folds are, more or less, coaxial to F_3 and are angular and often of kink band type accompanied by dominantly steep crenulation cleavage. Saklani and Bhatt (1989) estimated the flattening of minor folds of the Dharkot Nappe and the F_3 folds show 88 % of flattening while it is 80 % in F_2 folds. They (1989) observed that the minor folds (F_3 and F_4) exposed near the Dharkot Thrust exhibit 75 % of flattening and away from it the folds witnessed up to 65 % of flattening.

The inner zone of the Chail in the Lesser Himalaya is characterized by argillocalc-arenite rocks (Chail) occurring as tectonic windows (*e.g.* Uttarkashi, Dunda, north of Tehri) which are bordered by the Dharkot and Pratapnagar Thrusts

The middle zone of the Garhwal Lesser Himalaya is mainly characterized by three SW - vergent nappes (*e.g.* the Chail Nappes of Fuchs and Sinha, 1978).

According to Fuchs and Sinha (1978) the rocks of the Chail Series represent an imbrication zone and occur in the form of three tectonic units: The Chail Nappe 1 (C_1) is mainly represented by the parautochthonous carbonate sequence with associated slates and ferruginous quartzites. This is bordered in North by the Pratapnagar (Berinag) Thrust and in South by the Dharkot (Srinagar, North Almora) Thrust and occurs in the form of a window/ half-window. The Pratapnagar (= Berinag /Chail) Quartzite is allochthonous and solely constitutes the Chail Nappe 2 (C_2) of Fuchs and Sinha (1978). With regard to the phyllite (Gwar, Pauri, Dharasu, Saryu) it may be remarked that it is delineated at its base by the Dharkot (Srinagar, Dharasu, North-Almora, Chail) Thrust and belongs to the Chail Nappe 3 (C_3) unit of Fuchs and Sinha (1978).

Krol Nappe (Fig. 6.1)

Narendranagar Syncline

The rocks of Krol Series are well exposed around Narendranagar in a synclinal form. The high grade Krol C Limestone constitutes the core of it and is successively followed by Krol B Dolomitic Limestone (with red and green shales), Infrakol Slates and Blaini Boulder Bed. At Hindolakhal, the drag folds in the Infrakrol Slates exhibit slaty cleavage. At Duadhar (near Hindolakhal), the pink ferruginous quartzite (Nagthat) tectonically rests over the brick-red and olive green nummulitic shales (Eocene) along the Krol Thrust.

The minor folds of Krol Series were mainly controlled by the Krol Thrust. The effects of strain were high in the proximity of the thrust .The finite strain witnessed by the quartz grains was 0.34 near the thrust and was 0.20 away from it.

Near the village Asena, about 20 km from old Tehri, an exposure of large fold structures in Pratapnagar (= Berinag) Quatzite are present. In SW of Tehri near Chamba, the Chandpur phyllites and quartzites show flat dip to NNE, but also variable towards W and SW. Fold fabrics appear rarely developed in them. Folds are found not far from Chamba and their hinges deviate from Himalayan trend and plunge towards N to NNE.

The major Krol Syncline constituted by massive limestone and red shales is exposed and is overstepped by the Chail Nappe 2 at the village Nagni (11 km southwest of Chamba and 34 km northeast of Narendranagar).

At the village Khari (27 km northeast of Narendranagar) near the waterfall, the light coloured Krol Limestone tectonically overlies the dark Chandpur Phyllites (C_3) due to folded thrust. In Simla Slates, typical chevron folds with widely spaced crenulation cleavage are present near village Jajal (23km northeast of Narendranagar) (Fig. 6.8). Similar fabrics are found in Chandpur Phyllites near Agra Khal (Fig. 6.9).

Further to SW, near the village Duadhar (9km north of Narendranagar) Simla Slates capped by red Subathu Shales (Eocene) are exposed as a window bordered by the Krol Nappe formation. This Phart Window (named alter the village Phart) of Subathu lies morphologically in a deep valley.

Along the road from Duadhar to Narendranagar, Nagthat Quartzite, Chandpur Phyllite-Quartzite (rhythmic bedding) with kink band folds (Fig. 6.10), Infra-Krol Slates, Krol Limestone (Krol C), red shales (Krol B) and Blaini Boulder Bed (glacio-marine) are exposed. In the quartzitic bands of the Infra-Krol perfectly developed folds are present at Narendranagar (Fig. 6.11).

Below the town Narendranagar, along the road towards Rishikesh, red and green Blaini Boulder Beds, Infra-Krol Slates and Quartzites with kink bands are present (Fig. 6.12). The Blaini Boulder Bed as well as Infra-Krol Slates and Marls are exposed due to folding. The traverse is terminated towards SSW below Narendranagar at Rishikesh where the MBT is concealed below the Holocene.

Mussoorie Syncline

This syncline is characterised by the Satengal klippe. Dubey and Jayangondaperumal (2005) suggested that the Main Boundary thrust (MBT) in this region is constituted by the two thrusts. The first one is characterized by the combination of frontal and oblique ramps. While second thrust was initially a blind thrust and later outlined the Satengal Klippe of the Mussoorie syncline (Fig. 6.14).

Dubey and Jayangondaperumal (2005) carried out two experiments. In the first experiment the multilayer fault did not extend into the upper clay slab. It was found that the displacement was variable along the frontal ramps (more with leading frontal ramp and less with trailing frontal ramp). The compression given was upto 12%. Suferposed folds were noticed on the hangingwall and footwall. Subsequently the model was compressed upto 17% and a horizonal slice was cut from the deformed hangingwall to expose the internal structures above the blind thrust. It was observed that the outcrop patterns were of closed type and partial klippen were associated with thrust.

The second experiment was carried out to study the nature of thrust displacement along leading and trailing frontal ramps. The blind thrust remained confined to the basement. When maximum compression was changed into horizontal direction the antiformal superposed folds were formed in the central parts with shortening of 23%. Inliers were noticed above the blind thrust and oblique ramp depended upon the amount of vertical component (Fig. 6.15a, b; 6.16).

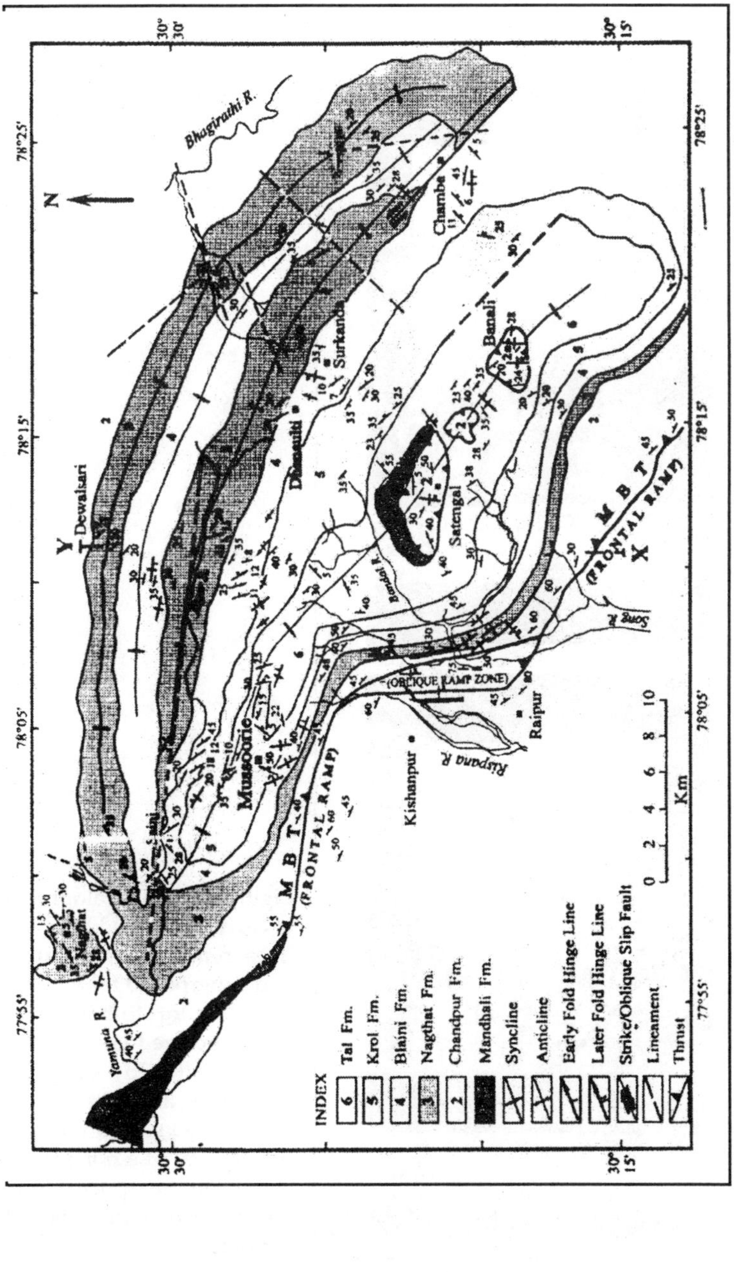

Figure 6.14: A geological and structural map of the Mussoorie Syncline, Lesser Himalaya (modified after Valdiya, 1980).

In figure the Oblique ramp is exposed in the field along the Song river while the Trailing frontal ramp is exposed northeast of Kishanapur. The Satengal and Banali klippens are also exposed (Fig. 6.14).

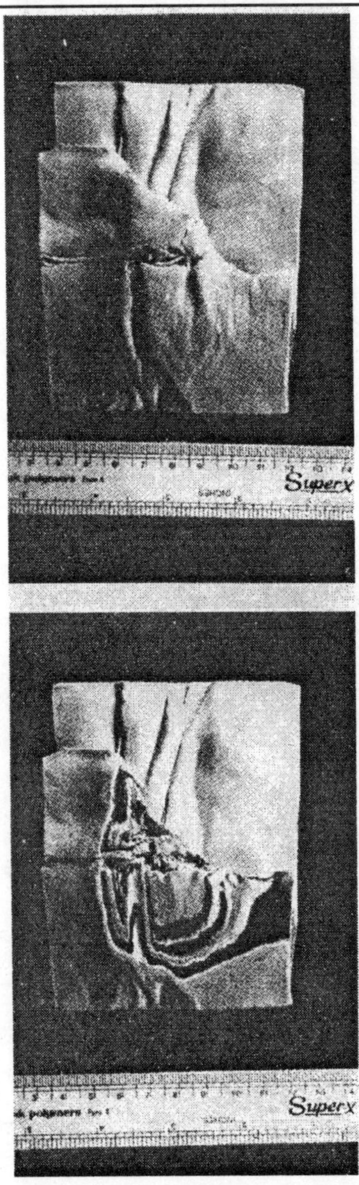

(a)

(b)

Figure 6.15a: Development of early folds (fold hinge lines trending left to right) and superposed folds (trending top to bottom) on a deformed layer surface (after 25% early shortening and 23% superposed shortening). **b.** Removal of a horizontal slice from the upper part of the model revealing partial klippen.

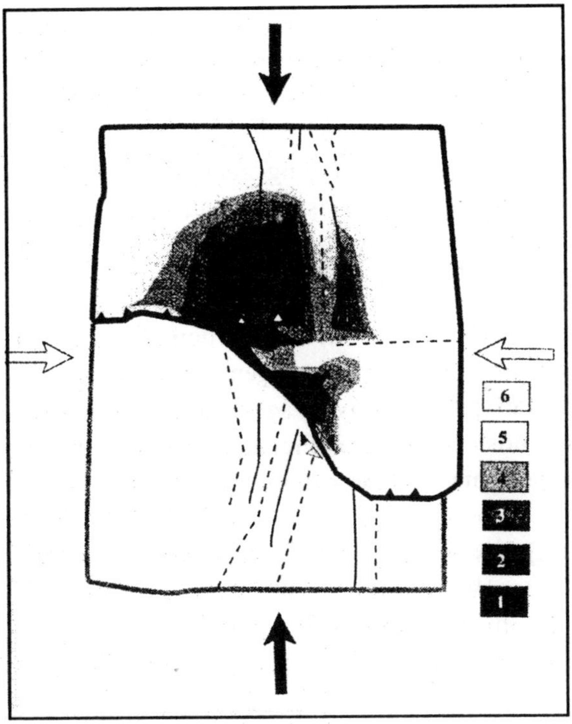

Figure 6.16: The structural features, 2 to 6 represent successively overlying layers in the multilayer packet. Solid arrows : axis of maximum compression during early deformation. Open arrows : axis of maximum compression (hinge lines).

Deformation History

Series of 5 generations of shortening structures B_1 to B_5 are taken into account for the reconstruction of the deformation history of the Garhwal Himalaya, These generations can be ascribed to the 5 deformational phases D_1 to D_5. Each generation has its own type of folds, cleavage planes, lineations, joints and thrust or fault planes.

Oldest mesoscopic compressional structures with lineations ($F_1 + F_2$ and $L_1 + L_2$) appear as cross or diagonal folds with cleavage ($S_1 + S_2$) with ± NNE trend in rocks of the Garhwal Himalaya. They are developed as upright, inclined or recumbent folds. The next generation is presented by wide-spread longitudinal ± NW striking structures showing strong SW-vergence and overturned, recumbent or plunging folds (F_3) with cleavage (S_3) and thrust planes (T_3). Thereafter, occur the WNW trending upright or inclined kink band folds and/or chevron folds (F_4). These are characterized by widely spaced, often steep dipping crenulation cleavage (S_4). They show weak SSW as well as NNE vergence. Compressional structures with a large wave length showing NE axial trends occur in large-scale anticlines and / or synclines (F_5).

The coordination of the five structure generations and/or deformation phases to specific orogenic episodes is not yet settled. A pre-Permian (pre-Himalayan) age for $F_1 + F_2$ is probable because true cross-folds occur only in Precambrian rocks of the Himalaya and corresponding structures in the basement rocks of the Indian Shield, *e.g.* in the Aravalli Mountains SW of Delhi (Auden 1937; Heim and Gansser 1939; Gansser 1964; Acharyya 1975; Valdiya 1976; Mehta 1977 and Schwan 1980, 1988). The F_3 and F_4 associated with S_3/T_3 and S_4/T_4 as well as the F_5 structures are mainly of Cenozoic (late-Alpine and/or Himalayan) age.

Saklani (1972) investigated the minor folds in the region of the Bhagirathi and Bhilangna valleys and described the first generation of folds (F_1) trending E-W; the second generation folds (F_2) plunge NW-SE while the third and fourth phase folds (F_3 and F_4) trend NE-SW and N-S. The tectonic modelof folding (Fig. 6.17) shows respective forces of folding (F_1 to F_4) as N-S, NE-SW and NW-SE to E-W.

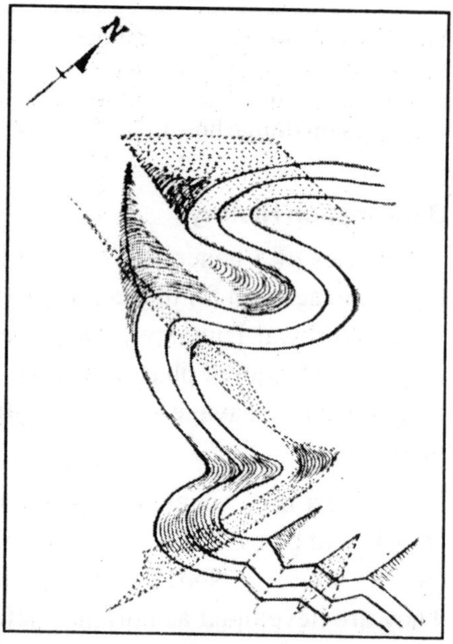

Figure 6.17: Teconic model showing the folding events.

References

Acharyya, S.K. 1975; Structure and stratigraphy of the Darjeeling Frontal Zone, Eastern Himalaya. —*Geol. Surv.Ind. Misc. Publ.,* **24** : 71-90; Calcutta.

Auden, J.B., 1937: Structure of the Himalaya in Garhwal. *Rec. Geol. Surv. Ind.,* **71**:407-433; Calcutta.

Dubey, A.K. and Jayangondaperumal, 2005, Satengal klippe in the Mussoorie syncline, Lesser Himalaya, Evidence from field and Model deformation studies, In : Saklani, P.S., 2005, *Himalaya (Geological Aspects)* **3**, 203-222.

Fuchs, G. and Sinha, A.K. 1978. The tectonics of the Garhwal - Kumaun Lesser Himalaya. *Jb. Geol. B.A.,* **121** : 219-241; Wien.

Gansser, A. 1964. *Geology of the Himalayas,* Intersci. Pub. : 269p. London.

Heim, A. and Gansser, A. 1939. Central Himalaya, Geological observations of the Swiss expedition 1936. *Dankschr.Schweiz, naturf. Ges.,* 73(1): 1-245; Zurich.

Mehta, P.K. 1977. Rb-Sr geochronology of the Kulu-Mandi belt: Its implications for the Himalayan tecto genesis.*Geol. Rdsch.,* **66**: 156-175; Stuttgart.

Ramsay,1981. Tectonics of the Helvetic Nappes. In : McClay, K.R. & Price, N.J. (Ed.): Thrust and Nappe Tectonics, *Spec. Publ. Geol. Society London,* **2**: 293-309; London.

Saklani, P. S. 1972. Lithostratigraphy and structure of the area between the Bhagirathi and Bhilangna Rivers, Garhwal Himalaya. *Himalayan Geol.,* **2**: 342-355; New Delhi.

Saklani, P. S. and Bhatt, S.C., 1989. Analysis of folds in Dharkot Thrust Sheet. In : Saklani, P.S. (Ed.); *Himalayan Thrusts and Associated Rocks,* **9**: 63-90, New Delhi.

Schwan, W. 1964. Begriff und Bedeutung der Leitstrukturen. Ein Beitrag zur tektonischen Forschungsmethodik. *Geotekton. Forsch.,* 19: 47 p. Stuttgart.

Schwan, W. 1973. Zur Ableitung der Grobtektonik eines Gebirges aus leitenden Strukturen senior Kleintekronik. Moglichkeiten und Grenzen einer geotektonischen Arbeitsmethode. In : Stockwerkbau und Felderteilung. *Veroff.Zentralinst. Physik der Erde,* **14(2)**: 317-352; Potsdam.

Schwan, W. 1980. Shortening structures in Eastern and North-western Himalayan rocks. In : Saklani P.S.(Ed.); 80 p.; New Delhi.

Schwan, W. 1981. Key structures of minor tectonics, major tectonic forms, and shortening kinematics in the Himalaya. In Sinha, A.K. (Ed.); *Contemporary Geoscientific Researches in Himalaya,* 1:13-20; Dehra Dun.

Schwan, W. 1988. Zur Gestaltung und Geschichte der Kompressionsstrukturen imIndischen Himalaya. Fakten und Fragen. *Erlanger Geol. Abh.,* **116**: 123-134; Erlangen.

Schwan, W. and Ray, K.K. 1987. Deviating shortening structures (B axes) in the Reyang Formation of the Iron Bridge area (Tista Valley, Darjeeling - Sikkim Himalaya / India). *N. Jb. Geol. Palaont. Abh.,* **174(3)**: 283-301, Stuttgart.

Schwan, W., Sinha, A.K. and Bagati, T.N. 1985. Structural tectonic observations along a traverse across the Chamba Region betweenNanikhad and Tisa in the Northwestern Himalaya, Himachal Pradesh, India. *N. Jb. Geol. Palaont. Abh.,* (9). 553-577. Stuttgart.

Schwan, W. and Saklani, P.S., 1991. Mesoscopic structures along the Bhatwari - Narendranagar traverse Garhwal Uttar Pradesh and their chronologic and kinematic integration of deformation plan of the Himalaya *Z. dt. Geol. Ges.,* 142, 97-113.

Valdiya, K. S. 1976. Himalayan transverse faults and folds and their parallelism with sub-surface structures of the North Indian Plains.- *Tectonophysics*, **32**: 353-386. Amsterdam.

Valdiya, K.S. 1980. *Geology of Kumaun Lesser Himalaya,* W.I.H.G., 291 p.

7

Structural Petrology

This branch includes microscopic structural features such as petrofabric analysis and petrotectonics. Identification of geometrical elements of rock fabrics are studied in petrofabrics while the relationship between the geometrical features and the movements responsible for the generation of the rock-fabrics and their causative stresses are included in petrotectonics.

Petrofabric Analysis

Sander (1910) was pioneer in the study of rock fabrics. The preferred orientation of minerals and their mutual relationship gives rise to a fabric-geometry and can be interpreted with reference to stress or stain.

Rock-fabric Elements

The shape of the grains, crystallographic planar or linear structures within the rocks are common fabric elements. The preferred orientation of minerals may or may not have a preferred lattice orientation consisting of crystallographic axes, optic axes, cleavage planes, translational twin glide planes or lines. The planar components of a fabric are called-S-planes which generally coincide with schistosity, slip, and shear planes. The linear components include c-axes of hornblende, quartz or the b-pinacoids of feldspar crystals. The crystallographic axes in a crystal, the planar and linear structures in a rock are referred to as a, b and c-axes which are mutually perpendicular to each other. The S-planes parallel to more than one axis of rock fabric are called as ab, ac and bc planes. The foliation plane in rock is parallel to a-axis while lineation or fold axis is parallel to b-axis and c-axis is perendicular to both.

Pole (or π) diagram of S-planes and linear (or β) diagram are used for fabric study. The plotting of these data is done on the lower hemisphere. This makes the observer to look the dip and plunge downwards as it normally does in geological fieldwork.

Orientation Maxima

There are mainly three types of maxima, namely, circular or point maxima, girdle maxima and cleft girdle or annular maxima.

In circular or point maxima, the orientation data cluster around a point while in girdle maxima the data show concentration along a great circle of the net. In the cleft girdle or annular maxima; the orientation data, are concentrated along a pair of opposite small circles of the net.

The point maxima show planar or linear fabric elements. The cleft girdle maxima are due to the presence of either linear or planar fabric element. Commonly, the rhombohedral calcite and quartz optic axes exhibit cleft girdles.

Petrotectonics

It deals with mechanism of deformation and internal movements withessed by the rocks. One can correlate the visible strain of the rock with those of individual movements of the fabrics and the stress system which affected the rocks. The rock fabric depends on the movements which affected the rocks. The direction of movements are reflected in the orientation of fabric data. The movements affecting the fabric elements are componental movements and these can be intergranular and intragranular. Intergranlar movements include rotation of the minerals and their sliding and preferred orientation. When the componental movements affect the crystal lattices of minerals then due to translation giding or twin gliding (Fig. 7.1, 7.2). The preferred orientation of the minerals takes place, and these are called as intragranular movements. External rotation may not cause preferred orientation of the lattice because axis of rotation and the maximum angle of rotation are not fixed. Due to gliding and dislocation the mineral grain can change its shape (Fig. 7.3). Considering the time relationship of deformation and crystallization it can be post-

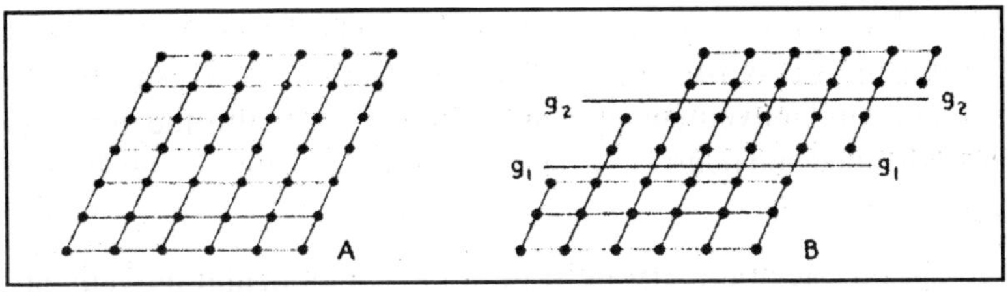

Figure 7.1: Translation gliding. Each dot represents center of a unit of the space lattice. (A) Before gliding. (B) After gliding along planes g_1g_1 and g_2g_2 (After Billings, 2000).

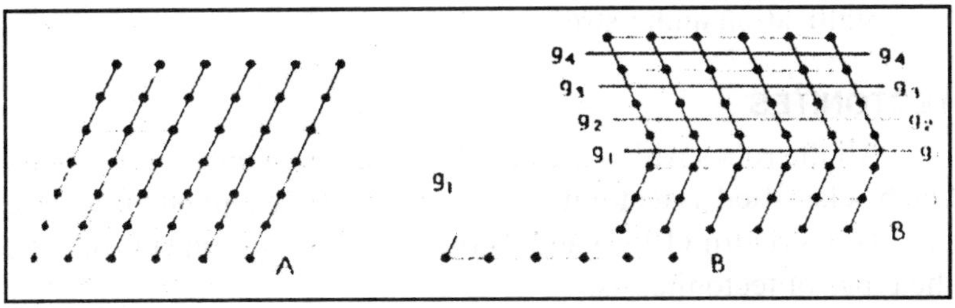

Figure 7.2: Twin gliding. Each dot represents center of unit of the space lattice. (A) Before gliding. (B) After gliding on planes g_1g_1, g_2g_2, g_3g_3 and g_4g_4. (After Billings, 2000).

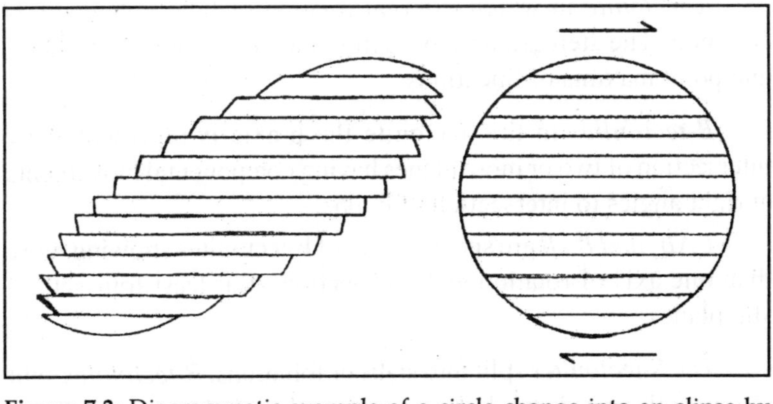

Figure 7.3: Diagrammatic example of a circle change into an elipse by movement along glide planes. (After Billings, 2000).

crystalline or pre-tectonic, pre-crystalline or post-tectonic and para-crystalline or syn-tectonic. In pre-tectonic crystallisation the componental movement are mainly direct. Due to this the minerals are deformed and quartz axes show deformation lamellae and micas get kinked.

In post-tectonic crystallization the componental movements are mainly indirect. The porphyroblasts are characterized by new structures (inclusions) and mimetic crystallization becomes effective. In syn-tectonic crystallization the direct and indirect componental movements are common. It promotes simaltaneous recrystallization under stress.

TECTONITES

All the rocks which are affected by componental movements can be classified as tectonites. *Sander* (1930). Knopf and Ingerson (1938), Fairbairn (1954) and Turner and Weiss (1963) proposed the types of tectonites as:

S-tectonite in which slip planes or foliation planes are prominent planar fabric elements and show single point maxima of poles.

R-tectonite in which external rotation of fabric elements is common. The sterograms show girdle maxima of poles of planes and point maxima of lineations.

B-tectonite: In this tectonite the b-axis is represented by intersection of two or more planes having componental movements at right angles to intersection of b-axis.

B∧B' and B⊥B are special types of b-tectonites showing more than one axis of rotation and intersection of at least four sets of slip planes.

The S-tectonites split into slabs or foliations, *R-tectonites* split into pencils or rods while *b-tectonites* split into lenses.

Symmetry of Fabrics

Sander (1925) remarked that symmetry of fabric and the movements are identical. There are four types of symmetry which can be axial or radial, orthorhombic, monoclinic and triclinic. Axial or rotational symmetry are rarely developed but can be observed in convolute folds and elongated pebbles in strongly compressed or streched rocks. Experimental studies have shown that random needles of rutile contained in a ball of clay get preferrentially oriented parallel to circumference of a cylinder formed by spreading (Fig 7.4).

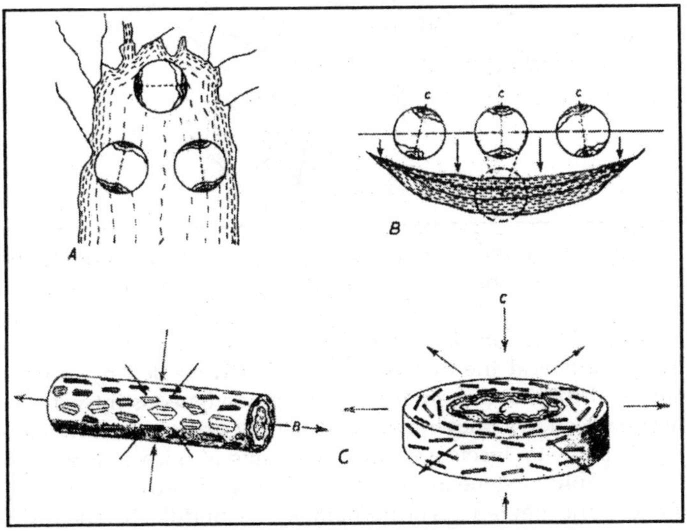

Figure 7.4: Axial Symmetry of Fabric in Diagrammatic Form

A. Vertical section of an igneous stock or cupola showing linear structure of hornblende prisms. Insets: local fabric diagrams of hornblende c. axes. Polar axial symmetry. (Modified from R. Balk (1936, After Hills, 1973).
B. Vertical section of lacustrine sedimentary basin showing direction of settling and ultimate orientation of flaky particles. Insets : local fabric diagrams of poles to clay flakes. Polar axial symmetry.
C. Central axial symmetry as expressed by the fabric of gympsum tablets in a rod produced by kneading (left) and of rutile needles in a disc obtained by 'spreading' of a ball of clay (right). Insets: integrated fabric diagrams. (Modified from W.O. Williamson (1955 After Hills, 1973).

Orthorhombic Symmetry: In this, the orientations are symmetrical with reference to three rectangular axes. The maxima can be divided into two parts along the deformation axes (Fig. 7.5).

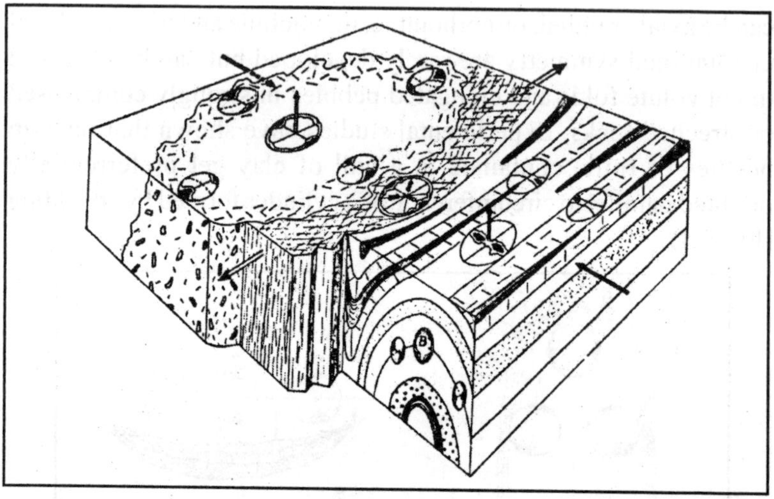

Figure 7.5: Orthorhombic Symmetry of Fabric in Diagrammatic Form Block-diagram showing, in the foreground, a symmetrical fold of the flexural slip type with fabric diagrams of mica cleavage poles as insets (monoclinic on the limbs, orthorhombic in the hinges). The fabric is inhomogeneous and the overall symmetry as revealed by integrated diagrams is orthorhombic. In the far distance an igneous intrusive body of oval cross-section is shown containing a flow structure with linear and planar components. Insets are fabric diagrams of mica cleavage poles. The fabric is inhomogeneous and the overall symmetry as revealed by integrated fabric diagrams is orthorhombic. The middle distance exhibits a strip of country rock flattened by slip on two equivalent and con jugate sets of planes intersecting in a vertical direction B. Fabric diagram (inset) of mica cleavage poles reveals total symmetry: three symmetry planes at right angles to each other, three twofold axes at their intersections : holohedral orthorhombic. Arrows represent major (← →) and minor (→ ←) deformation axes of the geological bodies concerned. Relative compression parallel to → ← is consistent with all three fabrics. B is the intersection of prominent sets of slip planes or the fold-axis (After Hills, 1973).

Monoclinic Symmetry: In this, there is only one plane of symmetry. The effective movement is along one set of slip planes parallel to *b* but inclined at various angles to the deformation axes (Fig. 7.6).

Triclinic Symmetry: In this the orientaion maxima can not be divided symmetrically. This is caused by steeply plunging folds in the nappes. (Fig. 7.7).

Figure 7.6: Monoclinic symmetry of fabric in Diagrammatic form. Cross-section of 'Gleitbrett' fold caused by slip on one set of planes. The planes are irregularly spaced and the slip does not vary regularly throughout the deformed area. A bed obliquely cutting the slip plane is shown in black. The undeformed central slice is a 'Gleitbrett; the upper portion is deformed by non-affine, the lower by affine slip on equally spaced slip planes. Insets; fabric diagrams of mica cleavage poles. Overall symmetry: hemihedral monoclinic; B is the normal to the only symmetry plane but is not, itself, a twofold axis of symmetry (After Hills, 1973).

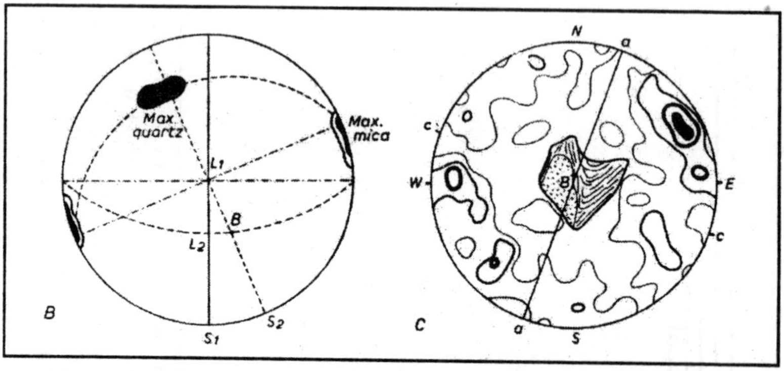

Figure 7.7: Triclinic Symmetry of Fabric in Diagrammatic form: B. Synoptic diagram of quartz optic axes and mica cleavage poles in a phyllite from Anglesey (England), showing principal maxima only. L1 and L2 are visible lineations, B is the (invisible) normal to the plane of monoclinic symmetry in the quartz fabric. S_1: original foliation; S_2 plane of transposition of mica flakes. (After Turner and Weiss, 1963). C. 446 optic axes of quartz from a vertically plunging fold in gneiss from the Cima Leinert, South Tyrol. Inset : oriented specimen from which section was made showing generalized trend of mica flakes. Contours: 5-4-3-2-1-0% per 1% area. Equal area projection. Lower hemisphere. (After O. Schmidegg, 1936 in Hills, 1973).

An arcuate fold belt in asymmetrical cross section and curvature shows triclinic symmetry of movement and fabric.

References

Billings, M.P., 2000. *Structural Geology* (3rd ed.), 606 p. Prentice Hall of India.

Fairbairn, H.W. 1954, *Structural Geology of Deformed Rocks*, Addison Wesley, Cambridge.

Hills, E.S., 1973. *Elements of Structural Geology*, 483 p., Methuen & Co. Ltd., London.

Knopf, E.B. and Ingerson, E., 1938, *Structural Petrology, Geol. Soc. Am. Mem.*, **6**, 68-72.

Sander, B., 1910, Uber Zusammenhange Zwischen Teilbewegung und gefuge in gesteinen, *Min. Pet. Mit.*, **30**, 281-314.

Sander, B., 1925, Zur petrographisch. - tektonischen Analyse II, *Jahr. Geol. Bundesanst, Vienna*, **75**, 181-236.

Sander, B., 1930, *Gefugekunde der Gestine*, Vienna.

Turner, F.J. and Weiss, L.E., 1963. *Structural Analysis of Metamorphic Tectonites*, 545 p. Mc Graw Hill Book Co., Inc.

• □□□ •

8

Petrotectonics of Tehri Garhwal

In Tehri Garhwal Himalaya, the Pratapnagar Quartzite is thrust over the parautochthon of limestone, slate and subordinate ferruginous quartzite of the Kotga Banali Group. With a view to examine the deformational trends in the Pratapnagar Quartzite in relation to the Pratapnagar Thrust, oriented samples of this rock were collected from near the thrust zone and also away from it. Thin sections containing 'ac' and 'bc' planes of the Pratapnagar Quartzite were analysed for petrofabric study.

The quartz exhibits various types of fabric orientations due to deformation. Fairbairn (1954) has classified the quartz axes orientations into eight types (Fig. 8.1). Saklani (1993) however, adopted only four generalized types of orientation, namely, maxima I, II, V and VII which are closely comparable with the orientations described by Fairbairn (1954). Maxima I of quartz axes orientations shows the single axes-concentration. These maxima are characterized by a spreading of axes in the form of a circular girdle. Maxima II occurs in two concentrations, more or less, symmetrically on either side of 'ab'. Maxima V shows small circle girdles enclosing 'c' and normal to 'ab'. Maxima VIII is characterized by concentrations enclosing 'b'. These orientation-types are, more or less, represented well in the Pratapnagar Quartzite.

In the fabric analysis of the Pratapnagar Quartzite the orientations of quartz axes in 'ac' plane have been rotated to 90° and superposed over the normal orientations of quartz axes in 'bc' plane with a view to know how far the relative positions of quartz axes of both the planes correspond with one another. The patterns of fabric deformation of the Pratpnagar Quartzite are shown in Fig. 8.2).

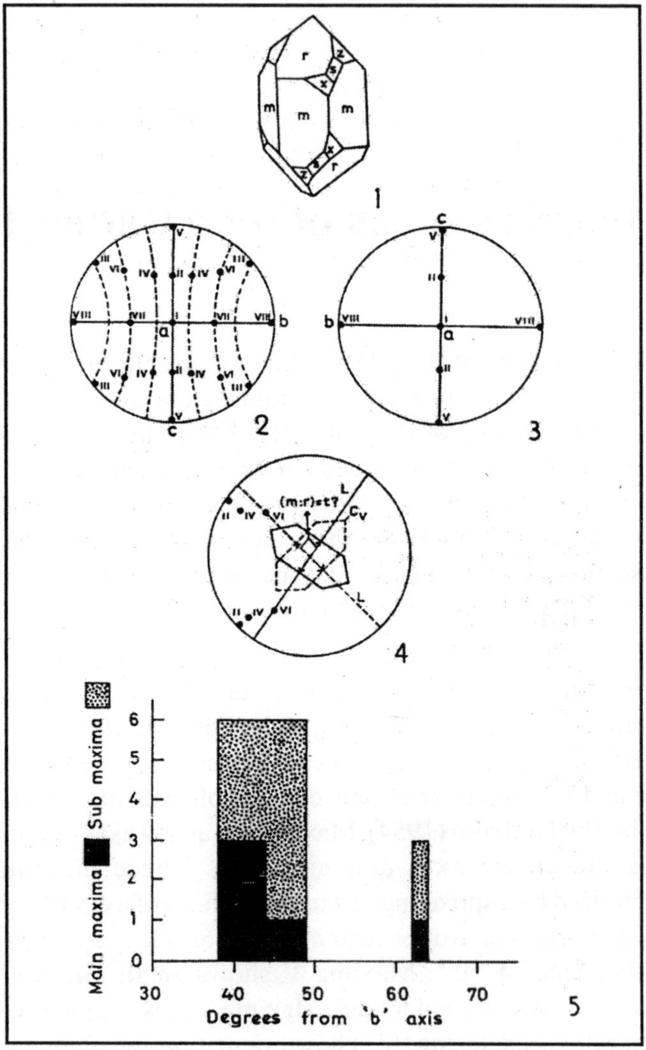

Figure 8.1: Showing the types of orientation of quartz axes in relation to quartz; translation gliding and graphical representation of orientation patterns of quartz axes in Pratapnagar Quartzite. 1. Crystal faces of quartz. 2. Relative position of quartz axes maxima (I to VIII), redrawn after Fairbairn (1954). 3. Important relative positions of maxima of quartz axes modified after Fairbairn (1954). 4. Translation glide redrawn after Fairbairn (1954). 5. Graphical representation of maxima and sub-maxima based on the combined diagram of 6 fabric diagrams of the Pratapnagar Quartzite. m = unit prism, r = positive unit rhombohedron, z = negative unit rhombohedron, l= lamella, t = assumed glide direction.

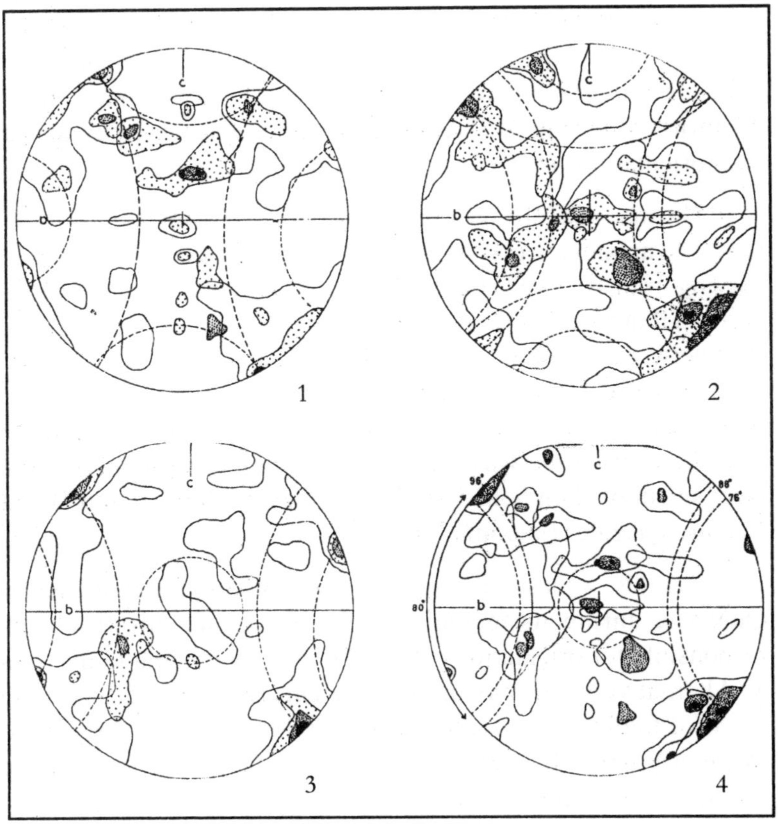

Figure 8.2: Showing contour-diagrams of fabric analysis.

1. Fabric diagram of the Pratapnagar Quartzite (near the Pratapnagar Thrust), containing normal 'bc' and rotated 'ac' sections of 140 quartz-axes plotted in the lower hemisphere. Contours 6%, 4% and 3% per 1% area.
2. Fabric diagram of the Pratapnagar Quartzite (near the Pratapnagar Thrust) containing 'bc' and rotated 'ac' sections of 299 quartz axes plotted in the lower hemisphere. Contours 6%, 4% and 3% per 1% area.
3. Fabric diagram of the Pratapnagar Quartzite (away from the Pratapnagar Thrust) containing normal 'bc' and rotated 'ac' sections of 140 quartz axes plotted in the lower hemisphere. Contours 6%, 4% and 3% per 1% area.
4. Combined diagram prepared from 24 maxima and 42 submaxima of 6 fabric diagrams of the Pratapnagar Quartzite containing 480 quartz axes.

I. Orientation pattern near the Pratapnagar Thrust:

Commonly, the Pratapnagar Quartzite near Pratapnagar and in between the Jakhnidhar and Bhelunta localities, situated along the thrust zone (Fig. 8.2). exhibits a fabric represented by quartz axes orientations belonging to maxima VIII, V and II or I.

The maxima VIII occurs from 40° to 125° with respect to ‘b’ axis. The maxima V, making a girdle around ‘c’, exhibits the angles varying from 40° to 92° from ‘b’. Sometimes, the quartz axes show maxima II in the quartzite near Pratapnagar while maxima I is shown by it between Jakhnidhar and Bhelunta localities. Maxima VIII, II and I are more dependent on rhombohedrons than those of maxima V. Therefore, it can be postulated that the Pratapnagar Quartzite near the thrust zone was subjected to deformation involving the mechanism of rhombohedral gliding (maxima VIII, II and I) and fractured needles (maxima V).

Possibly, a plastoviscous flow originated in the Pratapnagar Quartzite during its tectonic emplacement and this flow was responsible for orientations of quartz axes belonging to maxima V. Alternatively, the maxima V represents remnants of one of the phase of folding. The occurrences of oblique girdles belonging to maxima V have also been described from the granite-gneisses by Hofmann and Atnassow (1968).

II. Orientation pattern away from the Pratapnagar Thrust:

It has been noticed that the Pratapnagar Quartzite is marked by a distinct change in its fabric near Kholgad situated away from the thrust zone. The orientations of quartz axes (Fig. 8.2) are mainly represented by maxima VIII and II or I.

An interesting feature is observed in connection with the maxima VIII i. e. , the quartz axes orientations occur at smaller angles (40° to 76°) in contrast to those of the Pratapnagar Ouartzite exposed near the thrust zone. Further, a complete absence of maxima V is also noticed in the Pratapnagar Quartzite exposed away from the thrust zone. Some of the quartz axes exhibit a tendency to cluster in and around ‘a’ representing the maxima II. The axial girdle of maxima II is not well defined by a dominant trend of quartz axes.

Possibly, the heterogeneous distribution of quartz axes in the Pratapnagar Quartzite at Kholgad locality, lying away from the thrust zone, developed at the expense of homogeneously stressed quartz grains (Turner and Weiss, 1963, p. 325). Maxima II or I also represent some heterogeneous trends of quartz axes developed during the deformation of the Pratapnagar Quartzite.

Some of the optic axes of quartz are inclined at about 30° to the mica-foliation and some are subparallel to the mica-foliation, in the Pratapnagar Quartzite. The inclined orientations of quartz axes in specimens of the Pratapnagar Quartzite from near the thrust zone and away indicated the trend of deformation controlled by translation of the rhombohedral-planes of quartz (Fig. 8.2). The sub-parallel orientations of the quartz axes are suggestive of second trend of deformation related with the parallel alignment of fractured needles of quartz.

A study of the combined fabric-diagram of Pratapnagar Quartzite (Fig. 8.1 and 8.2) revealed the concentrations of quartz-axes gridle around 'b' at angles varying from 76° to 125°. Near the Pratapnagar Thrust Zone, the quartz axes ('bc' girdles) were oriented at high angles (76°, 88°, 96° to 125°) but away from it comparatively smaller angles were found upto 76° from 'b'.

The spread of quartz axes ('ac' girdles) around 'a' was exhibited particularly by the Pratapnagar Quartzite lying away from the thrust zone. Such 'ac' patterns are attributable due to local stresses produced at the expense of the synchronous deformation of the Pratapnagar Quartzite.

Thus, it is evident that the mechanism of deformation in the Pratapnagar Quartzite belongs to two different phases confined (i) along the thrust zone, and (ii) away from it. During the first phase, the Pratapnagar Quartzite has undergone deformation related to rhombohedral gliding of the quartz crystals syntectonically. However, the second phase of deformation in the Pratapnagar Quartzite along the thrust zone was controlled mainly by the fractured needles. Its conspicuous existence is amply supported by the presence of the 'c' girdle orientation of the quartz axes exhibited by the Pratapnagar Quartzite from the thrust zone only.

Strain Transition and Microstructural-Fabric analysis of Quartz -Mylonites exposed within Pratapnagar Nappe, Garhwal Himalaya was described by Bhatt and Saklani (1990).

The investigated area was in the Uttarkashi district of Lower Himalaya, consisting of Precambrian metasediments of Argillocalc-arenite Group, the Pratapnagar Group and the Dharmandal Group (Fig. 8.3). These units belong to the Chandpur Group of Auden (1937) and also to the Chail Group of Pilgrim and West (1928) in the Himachal Himalaya. Fuchs and Sinha (1978) held the view that the Chail rocks occur in the form of a composite nappe, i.e. Chail Nappes 1,2 and 3. The Pratapnagar Group corresponds to Chail Nappe 2 while the Dharmandal Group represents the Chail Nappe 3. The Argillocalc-arenite Group does not occur as a true nappe and is, in fact, parautochthonous. The Quartzite and the Phyllite Groups are metamorphics of greenschist facies (Fuchs, 1983; Windley, 1983). The Singuni limestone of Jain 1975 (= Deoban limestone) of the Argillocalc- arenite Group towards its eastern adjoining parts in Grahwal-Kumaun have yielded stromatolites (*Collenia baicalica and Collenia columnaris*) of Middle Riphean age (1000 to 1260 Ma- Valdiya 1969, 1980 and Saklani 1971). The Pratapnagar (Chail) Quartzite has an age of 1232± 120 Ma (Bhalla and Gupta, 1979).

A systematic sampling of the Pratapnagar Quartzite was carried out with reference to the thrust (Fig. 8.3). Five specimens were collected near the base of the thrust (PN 1- PN 5) and three samples were collected (PN6-PN8) from locations farther away from the Pratapnagar thrust (1250 to 1500 m).

MICROSTRUCTURES AND STRAIN DISTRIBUTION

The sections of rock were cut parallel to XY, YZ and XZ planes of finite strain ellipsoid where X, maximum principal axis of strain-ellipse is parallel to mineral lineation (Lm) and XY plane is parallel to mylonitic foliation (S_3); X,Y, and Z are orthogonal axes of an elongated quartz porphyroclast.

The plastically deformed quartzite consists of earlier formed host and later formed recrystallized quartz grains (Bhatt and Saklani, 1990). The host (old) quartz grains are globular, elongated/tabular and ribbon shaped (0.5 mm to 5mm). The recrystallized grains are equant or elongated in shape. The globular grains (0.20 mm to 0.40 mm) with an aspect ratio 2:1 are characterized by lobate boundaries and weak undulose extinction. The ribbon shaped grains (0.90mm to 1.50 mm) with an aspect ratio of 10:1 and the elongated/ tabular grains (0.50 mm to 1.00 mm-aspect ratio 5:1) are characterized by deformation bands and undulose extinction.

The tectonic strain was mainly accumulated in the deformed host (old) quartz grains in the Pratapnagar Thrust Sheet. The majority of the new recrystallized (equant) grains are strain free. A few of such grains (elongated) present at the margins of the elongated and ribbon shaped host (old) quartz grains are strained. Due to paucity and restricted occurrence such grains were not suitable for strain estimation. Therefore, only strained host (old) quartz grains having elongated shape were utilized for strain estimation. Earlier Marjoribanks (1976), Miller and Christie (1981), Law et al. (1984), Law (1986 and 1987), Bouchez and Pecher (1981), Bhatt and Saklani (1990) used such type of elongated host (old) grains as strain markers. All the different types of host (old) quartz grains were used for C-axes measurement.

The Pratapnagar Quartzite is mylonitized near the thrust (PN 1, PN2 and PN 3). It can be classified into type I S-C mylonite of Lister and Snoke (1984) and Law et al. (1984). In the type I S-C mylonite the S-surfaces (S_3) are defined by stretched or elongated quartz grains and the C- surfaces represent shear bands defined by phyllosilicates. Burg (1986), Davis et al. (1987), Culshaw (1987), Law and Potts (1987) have remarked that the oblique relationship between the mylonitic foliation (S_3) and the shear surface (C) is ascribed it to non- coaxial deformation.

In order to determine two dimensional strain the Rf and Ø parameters were measured in XZ, XY and YZ sections of each of the specimens and subsequently these values were matched with the strain ratio curves of Lisle (1985). The two dimensional strain ratio (Rs) in the Pratapnagar Quartzite exposed at the proximity of the thrust (Fig. 8.4), varied from 2.6 (XZ section of PN 3) to 3.2 (XZ section of PN 1) whereas it was the least in XZ section of PN8 (Rs = 1.70). The values of two dimensional finite strain ratio for

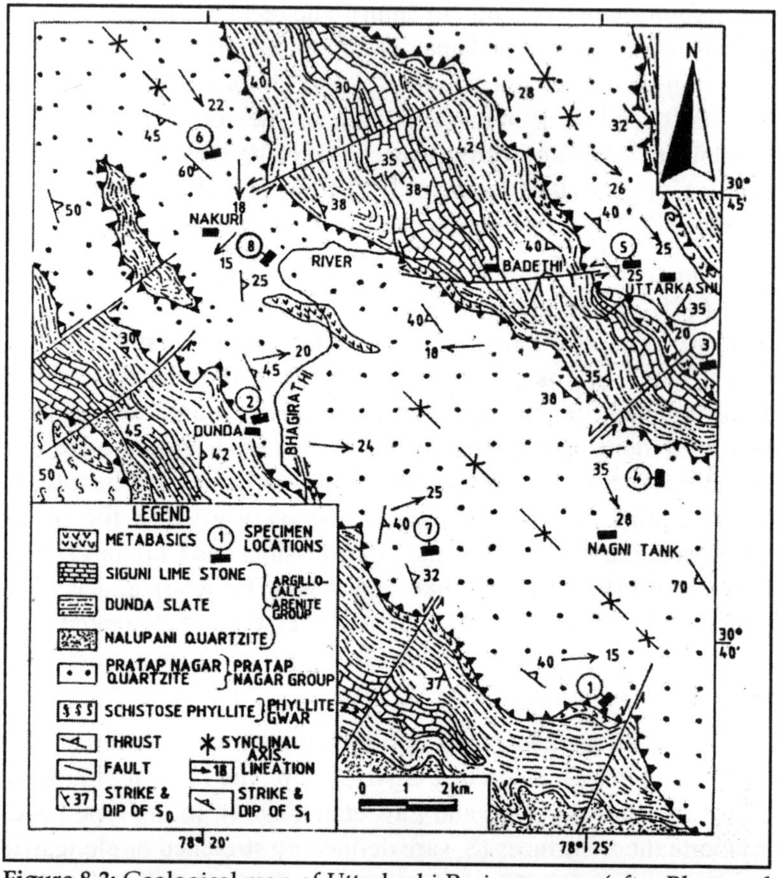

Figure 8.3: Geological map of Uttarkashi-Raripass area, (after Bhatt and Saklani 1990), Garhwal (Lower) Himalaya, showing specimen locations of Pratapnagar Quartzite.

XY and YZ sections were plotted on Flinn's graph and the shape (k) of these ellipsoids was determined (Fig. 8.4). The values of k parameters were also deduced from the equation as given by Ramsay and Huber (1983, p.200). The logarithmic strain parameters ($\Sigma 1$, $\Sigma 2$ and $\Sigma 3$) have been estimated by using the formula mentioned by Ramsay and Huber (1983; p.200). The three dimensional finite strain (Σs) estimated from the quartzitic specimens collected at the vicinity of the thrust is $\Sigma s = 0.0937$ (Bhatt and Saklani 1990). The quartzite exposed far away from the thrust (1500m) is characterized by low value of finite strain ($\Sigma s = 0.412$; Table I.).

Table I: Strain and Petrofabric (C-axes) data analysed within Pratapnagar thrust sheet.

Specimen Number	RXY (a)	RYZ (b)	RXZ (c)	Shape of strain elipse k	Angle between foliation and lineation (Lm)		Finite strain $\bar{\Sigma}s$	Distances from the thrust in meters
PN1	1.03	3.1	3.2	0.014	25^0	S3 & Lm (=x)	0.937	200m
PN2	1.07	2.6	2.8	0.043	22^0		0.812	300m
PN3	1.04	2.5	2.6	0.026	20^0		0.765	400m
PN4	1.25	2.0	2.5	0.25	18^0		0.676	500m
PN5	1.27	1.8	2.3	0.337	10^0		0.605	1000m
PN6	1.17	1.7	2.0	0.242	5^0	S1 & Lm	0.513	1250m
PN7	1.56	1.6	2.5	0.933	10^0		0.648	1300m
PN8	1.06	1.6	1.7	0.10	8^0		0.412	1500m

It is observed that the finite strain, degree of mylonitization, ductility and recrystallisation increase towards the thrust. The fabrics of ductile shear zones are generally deformed by pressure solution or crystal deformation. This process involved the transformation of material along grain boundaries and the fluid film that existed along these grain boundaries to facilitate the movement of dislocation through crystals and effect change in shape and volume of the grains (Price and Cosgrove 1990). The strain ellipsoids were found to be of oblate type and fell in the field of apparent flattening (Fig. 8.4). It was inferred that the homogeneous flattening strain and volume loss was prevalent in this shear zone. During homogeneous flattening strain (volume loss) the maximum elongation was parallel to X axis whereas the maximum contraction (shortening) was observed parallel to Z axis. But the computed values of k (Table I) and the ratio of strain axis (X:Y:Z), decreasing towards the vicinity of thrust (PN8-PN1) implied that the maximum elongation was initially parallel to X which subsequently became maximum parallel to intermediate Y axis

$$\{(1 + e_2)2 > (1 + e_1)\} + (1+e_3) \text{ of a strain ellipsoid.}$$

Thus, it was inferred that the mylonitized quartzite was deformed by high degree of homogeneous flattening type of finite strain at the vicinity of ductile Pratapnagar/Chail Thrust while away from the thrust it experienced low magnitude of flattening type of finite strain.

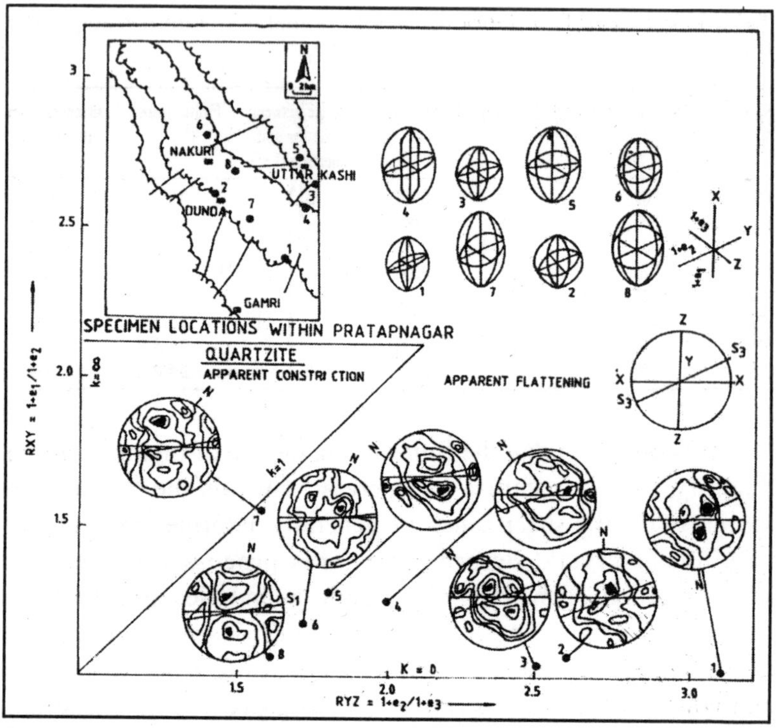

Figure 8.4: Flinn plot illustrating relationships between C-axes fabrics and estimated strain of host old quartz grains within Pratapnagar Thrust Sheet. The inset geotectonic map shows the specimen locations of Pratapnagar Quartzite. All the specimens are in the apparent flattening field showing oblate shape of strain ellipsoids.

C-AXES ORIENTATION OF QUARTZ FABRICS

The fabric diagrams of PN1 and PN2 from the immediate vicinity of the Pratapnagar Thrust (200-300m) were characterized by NNW-SSE trending typical oblique girdles (unpopulated inclined single girdle pattern of Lister and Hobbs, 1980), inclined at 35° to 40° with reference to X and the maxima centered about Y (Fig. 8.5a). The strike of mylonitic foliation (S_3) and trend of lineation (Lm = x) represented in these plots were from N55° W to N72° W and N60° W to N50° W respectively. The typical (Mancktelow, 1987a,b) N-S trending C-axes single girdles shown by stereographic plots of PN 3, PN4 and PN5 lying at 35° to 40° with reference to X demonstrated asymmetrical alignment of C-axes maxima with respect to mylonitic foliation (S_3) (Fig. 8.5a). These plots also exhibited variable orientations of mylonitic foliation (N65° W to N35° W) and lineation (N45° to N 25° W). The inclined alignment of the C-axes single girdle pattern with respect to mylonitic foliation and their subperpendicular relationship with mineral lineation (Lm) led to the inference that the C-axes fabrics were asymmetrically oriented with respect to mylonitic foliation (S_3).

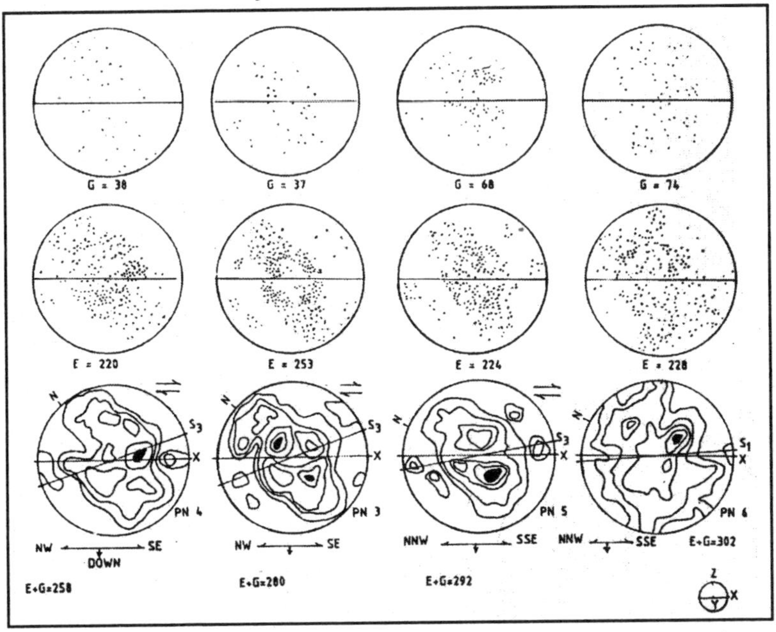

Figure8.5a: Contoured plots of C-axes for globular (G) and elongated (E) host (old) quartz grains, contour in 1-2-6-12% per 1% area (PN4), 1-2-8-14% per 1% area (PN3), 1-2-6-10-14% per 1% area (PN5) and 1-2-4-8 & > 8% per 1% area (PN6) of a net.

The NNW-SSE trending C-axes cross girdle (PN6) shows symmetrical distribution of C-axes fabric with respect to foliation. The quartzite exposed at 1300m (PN7) and 1500m (PN8) from the Pratapnagar Thrust displayed symmetrical corss and small circle girdle patterns of C-axes in which the maxima were located at 40° to 45° from Z (Fig. 8.5b). The contoured fabric plots of specimens PN6, and PN8 displaying the concentration of C-axes fabrics about Z and their shortening about X indicated axial shortening of fabrics

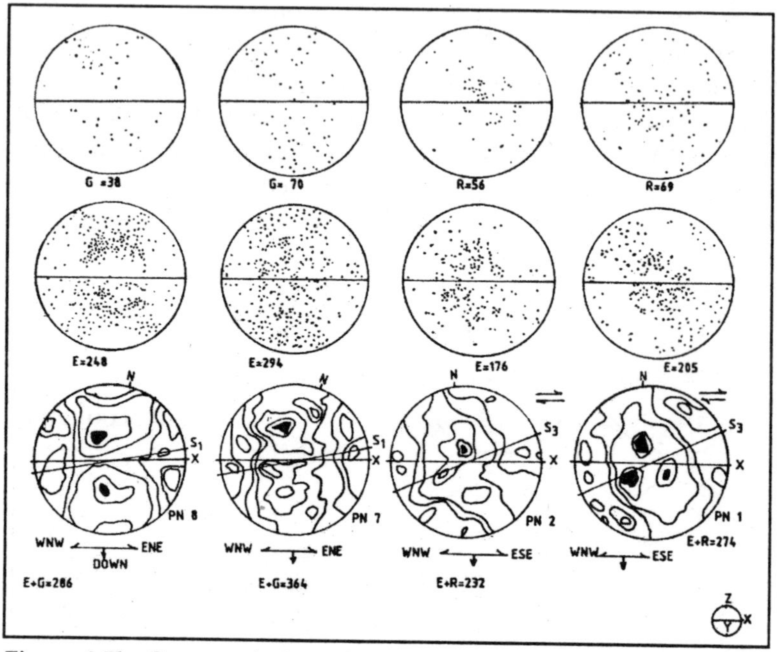

Figure 8.5b: Contoured plots of C-axes for globular (G), elongated (E) and ribbon (R) host (old) quartz grains contour in 1-2-8-14% per 1% area (PN8),1-2-4-8-12% per 1% area (PN7), 1-2-4-8 & > 8% per 1% area (PN2) and 1-2-6-10% per 1% area (PN1) of a net.

(Lister et al., 1978) in the flattening finite strain field (Fig.8.4). In fabric plot of PN7, the C-axes fabrics also centered about Z but the polar regions of the Z and X axes were not free and hence they were possibly produced due to progressive plane strain (k = 0.933).

The quartzite exposed at the proximity of the thrust zone was characterized by strong development of mylonitic foliation, S_3 (represented as s-surfaces), oblique shear planes (c-surfaces) and associated penetrative mineral lineation (Lm).

The kinematics of microstructures, strain indicators and C-axes fabrics examined within the quartzitic specimens belonging to shear zone revealed that the transposition of shear planes, rotation of porphyroclasts and sense of shear were generally governed by strain transition, mylonitization and recrystallization of quartz grains. The quartz porphyroclasts were rotated due to variable tectonic stresses and elongated under the influence of ductile flow.

As a result the new planar components (mylonitic foliation), and shear planes (c- surfaces) were formed.

The variable orientations of mylonitic foliation (S_3) and mineral lineation (Lm) plotted for each specimen revealed convincing macroscopic evidences for shear movement. It was observed that the mylonitic foliation (S_3) was changed in its strike from N35° W (PN 5) to N85°W (PN 1). In other words, it can be stated that the strike of the mylonitic foliation (S_3) was rotated sinistrally 50° from its original position (NNW-SSE). The associated stretching mineral lineation (Lm) also changed in its trend (N 25° W of PN 5 and N60° W of PN 1) due to effects of rotational movement. This sinistral vergence also brought obliquity between dip directions of mylonitic foliation (S_3) and plunge of lineation which varied from 10° (PN5) to 25° (PN1) along Pratapnagar thrust zone (Fig. 8.5 a, b).

The fabric diagrams of specimens (PN1-PN5) collected in the vicinity of Pratapnagar Thrust showed asymmetrical C-axes girdles inclined to mineral lineation (Lm) and mylonitic foliation (S_3). The NNE-SSW trend of asymmetrical C-axes single girdle (PN5) changed to NNW-SSE (PN 1 and PN 2). Thus it was inferred that these girdles rotated about 50° from N 25° E to N 25° W. The microstructural criteria for deduction of sense of shear were also observed within XZ sections of mylonitized quartzite. The two sets of planar components represented S- (S_3 mylonitic foliation) and C-(shear) planes. The C- surfaces inclined at 30° to 40° to the S-planes were approximately parallel to shear direction and strike parallel or sub-parallel to the trend of mineral lineation. This angular relationship between these planes also exhibited sinistral vergence of shear. The quartz porphyroclast with asymmetrical,pressure shadow also indicated the sinistral vorticity of shear movement. However, the rotation of the fabric was rather insignificant away from the thrust. This phenomenon is shown by randomly oriented globular and tabular quartz host grains and symmetrical orientation of C-axes girdles (PN6, PN7 and PN8).

The presence of microstructural evidences, viz; elongated and ribbon shaped quartz grains, preferred orientation of S- and C-planes, obliquity between mylonitic foliation (S_3) and mineral lineation and asymmetrical orientation of inclined C-axes girdles with respect to foliation and lineation proved the existence of progressive simple shear (non-coaxial) deformation along the Pratapnagar Thrust zone. Away from it the occurrence of cross and small circle girdles, presence of randomly oriented globular and tabular grains and lack of meso-and microstructural shear criteria suggested that the quartzite was affected by coaxial deformation.

The mylonitization accompanied by dynamic recrystallization became intensive in the vicinity of the thrust plane. As a result the proportion of quartz porphyroclast was reduced and the recrystallized strain free quartz grains (equant) were formed. Such a type of mylonitic rock is termed as ultramylonite.

Thus, the mylonitized quartzite was subjected to continuous effects of deformation under ductile and low grade metamorphic conditions. Due to rotation of tectonic stress directions the stretched fabrics were also oriented subparallel to thrust transport (NE- SW). The occurrence of the oblique NE-SW trending transverse faults indicating counterclockwise vergence suggested that the ductile Pratapnagar/Chail Quartzite was deformed by brittle (cataclasis) shearing in a later stage.

References

Auden, J.B. 1937. Structure of the Himalaya in Garhwal, *Rec. Geol. Surv. India,* **71**, 407-433.

Bhalla, N.S. and Gupta, J.N. 1979. U-Pb Isotopic ages of Uraninites from Kulu, Himachal Pradesh and Berinag, Uttarpradesh, *Jour. Geol. Soc. India,* **20** (10), 481-488.

Bhatt, S.C. and Saklani, P.S., 1990. Kinematic framework of heterogeneous deformation within Pratapnagar thrust sheet, Bhagirathi Valley, in Lesser Garhwal Himalaya, U.P., *Jour. Geol. Soc. India,* **36** (3), 247-261.

Bouchez, J.L. and Pecher, A., 1981. The Himalayan Main Central Thrust pile and its quartz-rich tectonites in Central Nepal, *Tectonophysics,* **78**, 23-50.

Burg, J.P., 1986. Quartz shape fabric variations and C-axes fabrics in a ribbon-mylonite: arguments for an oscillating foliation, *Jour. Struct. Geol.* **8**(2), 123-131.

Culshaw, N., 1987. C-axis pattern microstrain and kinematics of some S-C mylonite in Greenville gneiss, *Jour. Struct. Geol.*, .**9**(3),.299-311.

Davis, G.H. Gardulski, A.F. and Lister, G.S., 1987. Shear Zone origin of quartzite mylonite and mylonitic pegmatite in the Coyote Mountains metamorphic core complex, Arizona, *Jour. Struct. Geol.,* **9**(3), 289-297.

Fairbairn, H.W., 1954. *Structural Geology of Deformed Rocks*, Cambridge - Rddison Wesley Press.

Fuchs, G., 1983. The Chail Thrust and Chail Nappes in the Himalaya. In : Saklani, P.S. (editor), *Himalayan Shears*, Himalayan Books, New Delhi, 33-44.

Fuchs, G. and Sinha, A.K. 1978. The Tectonics of the Garhwal Kumaun Lesser Himalaya, *Jb. Geol. B-A.,* **121**(2). 219-241.

Hofmann, J. and Atnassow, O., 1968. Der Rila Pluton and Sein Ostlicher Rahmen - ein Beispiel Zur magme taktonischen Bearbeitung granitischer korper, *Mon. Deutch. Akad. Wissen* **10**, 704-717.

Jain, A.K., 1975. Structure and Petrology of Mylonite and Related Rocks from the Lesser Himalaya, Garhwal, India, *Geol. Rund.* **64**(1) 230-248.

Law, R.D., 1986. Relationships between strain and quartz crystallographic fabrics in the Roche Maurice quartzites of Plougastel, Western Brittany, *Jour. Struc. Geol.,* **8**(5) 493- 515.

Law R.D. 1987. Heterogeneous deformation and quartz crystallographic fabric transitions: natural examples from the Moine Thrust Zone at the Stack of Glencoul, northern Assynt, *Jour. Struct. Geol.,* **9**(7), 819-833.

Law, R.D., Knipe, R.J. and Dayan. H. 1984. Strain path partitioning within thrust sheets: microstructural and petrofabric evidence from the Moine.Thrust Zone at Loch Eriboll, north-west Scotland, *Jour. Struct. Geol.*, **6**(5) 477-497.

Law, R.D. and Potts, G.J. 1987. The Tarskavaig Nappe of Skye, north-west Scotland: a re-examination of the fabrics and their kinematic significance, *Geol. Mag.*, **24**(3), 231-248.

Lisle, R.J. 1985 Geological Strain Analysis, a Manual for the Rf = Ø Method, Oxford Pergamon Press, pp.99.

Lister, G.S., Paterson, M.S. and Hobbs, B.E. 1978. The Simulation of fabric development in plastic deformation and its application to quartzite: the model, *Tectonophysics*, **45**, 107-158.

Lister, G.S. and Hobbs, B.E. 1980. The simulation of fabric development during plastic deformation and its application to quartzite: the influence of deformation history, *Jour. Struct. Geol.,* **2**(3), 355-370.

Lister, G.S. and Snoke, A.W., 1984. S-C Mylonites, *Jour. Struct. Geol.,* **6**(6), 617-638.

Marjoribanks, R.W., 1976 The relation between Microfabric and strain in a progressively deformed quartzite sequence from Central Australia, *Tectonophysics*, **32**, 269-293.

Miller, D.M. and Christie, J.M., 1981. Comparison on quartz microfabric with strain in a recrystallized quartzite, *Jour. Struct. Geol.*, **3**, 129-142.

Pilgrim, G.E. and West, W.D., 1928. The structure and correlation of the Simla rocks, *Mem. Geol. Surv. Ind.* **53**, 1-140.

Price, N.J. and Cosgrove, J.W., 1990. *Analysis of Geological Structures*, Cambridge University Press, 502 p.

Ramsay, J.G. and Huber, M.I. 1983. The *Techniques of Modern Structural Geology*, **I**: *Strain Analysis, Academic Press*, London, 307 p.

Saklani, P.S. 1971. Occurrence of Stromatolites in the Bhelunta Limestome formation of Pratapnagar area, Garhwal Himalaya, *Science and Culture*, **37**, 215-216.

Saklani, P.S., 1993. *Geology of Lower Himalaya (Garhwal),* 240 p. International Books, Delhi.

Turner, F.J. and Weiss, L.E., 1963. *Structural Analysis of Metamorphic Tectonites*, McGraw-Hill, New York,.560 p.

Valdiya, K.S. 1969. Stromatolites of the Lesser Himalayan Carbonate formations and the Vindhyans, *Jour. Geol. Soc. Ind.*, 10, 1-25.

Valdiya, K.S. 1980. *Geology of Kumaun Lesser Himalaya,* W.I.H.G., Dehradun. 291 p.

Windley, B.F. 1983. Metamorphism and tectonics of the Himalaya, *Jour. Geol. Soc. Lond.*, **140**,.849-865.

•□□□•

9

The Himalaya and the Alps

The Himalaya has been frequently compared with the Alps (Gansser, 1964 b., Valdiya 1964; Krishnaswamy and Swaminath 1965; Wenk 1968). The tectonic movements in the Himalaya remained active from north, producing a vergence towards the south, while those in the Alps were from south and produced a northward vergence. The history of the Himalaya began in the Cambrian times and is exemplified by the occurrence of the Salt range in Pakistan. Salt might have acted as a lubricant for the thrusting of the rocks. That is why the Salt range has also been described to be of thrust-controlled origin. The Alpine history began in the Triassic and was then characterized by the occurrences of evaporites (Gansser, 1964 a). Similar tectonic significance of the salt can also be observed in the Jura series of the Alps. However, this phenomenon is not noticed in the Pennine and Lepontic Alps. The Himalayan foreland, constituted by the Siwalik, is exposed at a few hundred metres above the mean sea level, whereas the molasse of the Alpine foreland is exposed at the heights of the middle Alps (Gansser, 1964 b). The Himalayan foreland is towards the south whereas the Alpine, foreland is towards the north. So far, the rocks affected by the Caledonian-Hercynian orogenesis have not been reported from the Himalaya, and therefore, the equivalents of the Monte Rosa, Aar-Gotthard, Aiguille-Rouges-Massifs of the Alps are not represented in the Himalaya (Figs 9.1, 9.2). However, the Jutogh, Garhwal, and Darjeeling nappes can be compared with the Pennine thrust sheet of the Alps (Heim, 1956). The syntaxial bends (Wadia, 1932) of the Himalaya are of tectonic origin; in the Alps they are due to the characteristics of the foreland (Krishnaswamy and Swaminath, 1965).

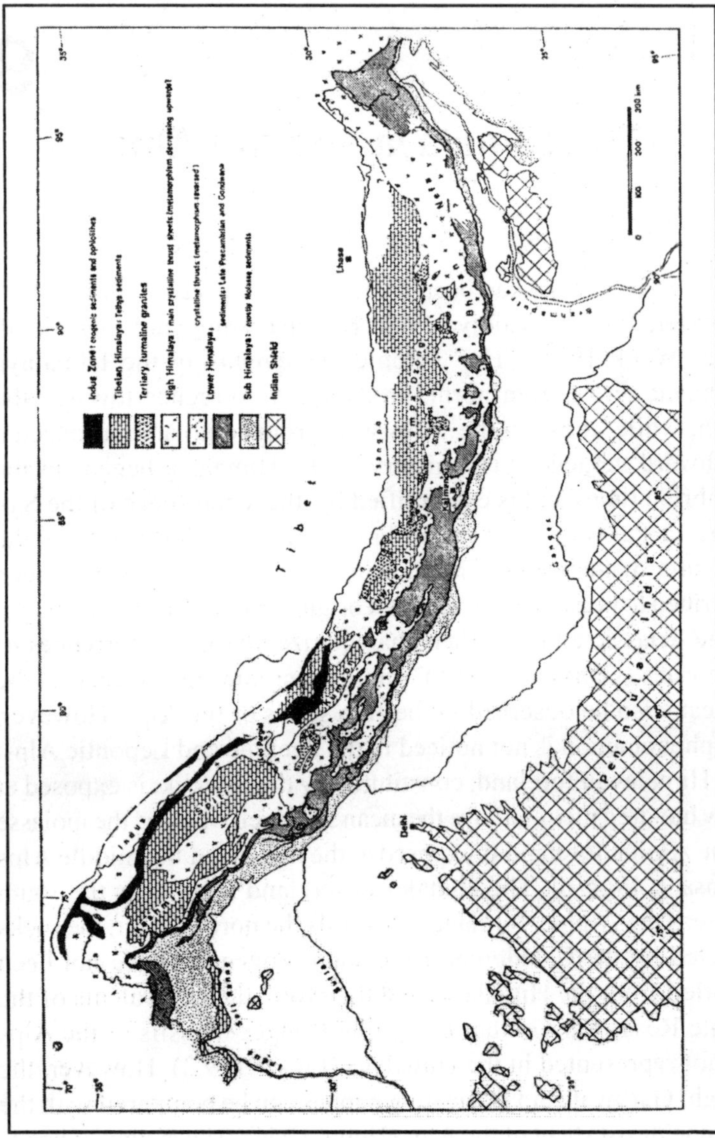

Figure 9.1: Generalized map of the Himalaya after Gansser (1973).

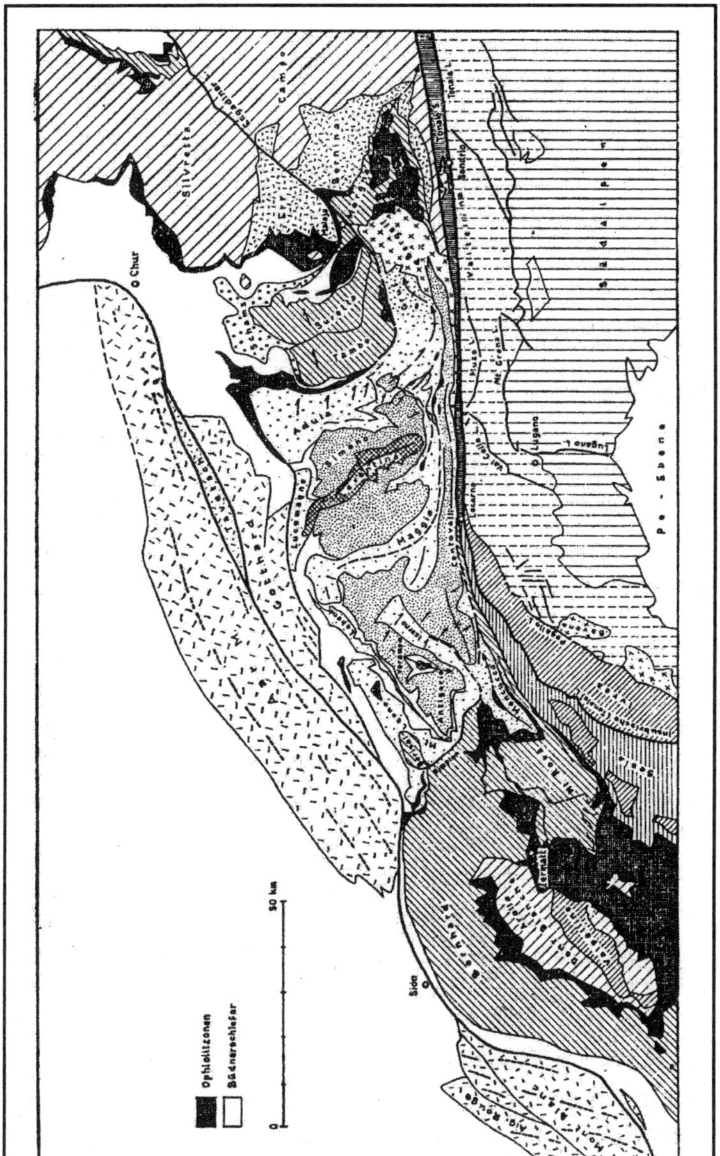

Figure 9.2: Generalized map of the Western Alps after Gansser (1973).

Similarities

Both the mountain chains are characterized by some similarities as follows:

(i) The presence of the molasse-like sediments, with associated system of reverse faults.

(ii) The occurrence of thrust sheets and erosionally controlled windows - and klippen structures (Valdiya, 1964).

(iii) The existence of the wrench-fault system at the roots of the main thrusts.

(iv) The occurrence of Tertiary granites in the Crystalline Zone (= Innsbruck zone of the Alps).

Dissimilarities

(i) The rock facies of the Himalaya are traceable with regularity whereas the Alpine rock facies are characterized by many irregularities (Krishnaswamy and Swaminath, 1965)

(ii) The flysch-like sediments are mostly absent in the Himalaya whereas these are well developed in the Alps (Figs 9.1 and 9.2).

(iii) There is a conspicuous absence of the "Bundner Schiefer" in the Himalaya in contrast to the Alps (Wenk, 1968).

(iv) Although the basement of the Himalaya is Precambrian yet it seems that it remained unaffected by the Caledonian/ Hercynian orogenesis. The occurrence of the Stromatolitic limeslone is significant in connection with the existence of the Precambriain rocks in the Himalaya.

(v) The molasse-like sediments of the Himalaya are of freshwater origin, while some molasse sediments in the Alps are marine (Kordiuk, 1938; Trumpy, 1960).

SMALL SCALE STRUCTURES

The structural and metamorphic studies in the last two decades have been of considerable interest in the Himalaya. However, there is a lack of the literature pertaining to a comparative study of the minor structures of the Himalaya and the Alps (Table I).

Table 1

	Author	Area	F 1	F 2	F 3	F 4
HIMALAYA	MERH (1966)	Kumaun	N to NE	WNW — ESE	N to NE	—
	NAHA & RAY (1970)	Simla	E — W	E — W	NNW — SSE	—
	BHATTACHARYA & NIYOGI (1971)	Simla	NW — SE	NW — SE	NW — SE	—
	SAKLANI (1971—1974)	Garhwal	E — W	NW — SE	N — S to NE — SW	N — S to NE — SW
ALPS	KVALE (1957)	Gotthard, Piora Scopi	N — S	E — W	—	—
	WUNDERLICH & PLESSMANN (1957—58)	Aar-Gotthard Lucomagno	N — S (B_1) variable also	E — W	—	—
	CHATTERJEE (1961)	Simplon	N — S (B_1)	E — W (B_2) ENE — WSW	—	—
	MILNES (1968—1973)	Simplon	N — S, SE — NW	NW — SE	—	—
	CHADWICK (1968)	Lukmanier	Nappe formation and phase B (N — S)	Phase V E — W,	—	—
	THAKUR (1973)	Molare	Nappe formation and N — S	E — W	—	—

1. Himalayan Minor Structures: A comparison of the minor structural elements is given in Table I. It is evident that differences of opinions exist with regard to the sequence of structural events in the Himalaya. According to Naha and Ray (1970) the E-W trending structures belong to first and second phases of deformation, but Bhattacharya and Niyogi (1971) regard the NW-SE-trending structures belonging to the first and second generations of deformation. The investigations by Merh (1966) in Kumaun and by Saklani (1971a,b, 1972 a,b, 1973 a,b; 1974) in Garhwal seem to be somewhat similar. However, Merh (1966) described the first phase structures showing a trend from N to NE, but according to Saklani (1971, 1972 a, b, 1973a, b) they were with E-W-trend in Garhwal. It is quite possible that the E-W-trending structures of Simla and Garhwal were influenced to a great extent by the subsequent movements of deformation and hence the originally E-W-trending structures were marked by a variable attitude from N to NE in the Kumaun Himalaya.

Saklani (1972 a, b, 19731,b) described a fourth type of deformation (coaxial to the third phase of folding) which developed near the Dharkot Thrust zone (= Garhwal Thrust), and this was attributed due to syn- to post-thrusting movements associated with brittle deformation. Since such a type of thrust-controlled deformation is well marked in Garhwal it is, therefore, expected that future studies would further elucidate this specific aspect ot deformation in the adjoining areas.

2. Alpine Minor Structures: There seems to be a general agreement with regard to the earliest deformation associated with thrust-gliding movements (Table I).

The first phase N-S-trending deformation was described in Ursen, Garvera, Nufnen, Piora Scopi areas by Kvale (1957), in the Aar-Gotthard and Lucamagno areas by Wunderlich (1958) and Plessmann (1957, 1958), in Simplon area by Chatterjee (1961), in Lukmanier area by Chadwick (1968) and in Molare area by Thakur (1973). Milnes (1968, 1973) has additionally remarked that the early Alpine lineation in the Simplon area showed variable orientation from N-S, SE to E-W. According to him the second phase- deformation (L_2) mainly trends NW-SE and not E-W as has been described by earlier workers. In the Eastern Alps (S-Radstadter Tauern), similar to the Simplon area, the axes of the Alpine main structures of the same age changed their orientations from, NW-SE to E-W to N-S, but there were more than four different lineations (acts of movement) (Schwan ,1965). It is, therefore, evident that there is a lack of unanimity about the sequence of events of the structural deformation in the Alps, but also primary difference in the orientations of the structures. It has been noticed that some of the linear structures present in the Aar-Gotthard and Pennine rocks were traced persistently in the enveloping Mesozoic cover and these were assigned to the last and final phase of Alpine deformation.

The comparison of the structures of the Himalaya and the Alps reveals that both the mountain chains have their own structural and tectonic characters Generally speaking, the mechanism responsible for the large scale structures is similar, but its effects in detail are quite different in the areas of the Himalaya and the Alps The studies of both the mountain chains are complementary to one another in understanding the structural complexities.

References

Bhattacharya, S.C., Niyogi D. 1971. Geologic evolution of the Krol Belt in Simla Hills, H.P., - *Himalayan Geology*, **1**, 178-212.

Chatterjee N.D.,1961. The Alpine Metamorphism in the Simplon Area, Switzerland and Italy. *Geol. Rdsch.*, **51**, 1-72.

Chadwick, B. 1968. Deformation and Metamorphism in Lukmanier region, Central Switzerland. - *Bull. Geol. Soc. Am.*, **79**, 1123-1150.

Gansser, A. 1964a. *Geology of the Himalayas* - Interscience Publishers John Wiley and Sons Ltd., 289 p.

Gansser, A. 1964b. The Alps and the Himalaya. *Proc. 22nd Int. Geol. Cong.*, **9**(2), 389-399, Delhi.

Gansser, A. 1973. Orogene Entwicklung in den Anden, in Himalaja und den Alpen, ein Vergleich. - *Elcogae Geologicae Helvetiae*, **66**(1),23-40.

Heim, A. 1956. The Geological Structure of the Himalaya compared with the Alps. - *Proc. Nat. Inst. Sci.Ind.*, 22 (A/4), Delhi.

Kordiuk, B. 1938. Zur Entwicklung des subalpinen Molassetroges. - Abh. Preuss. Geol. L. - A., N. F.,

Krishnaswamy, V.S., Swaminath J. 1965. Himalayan and Alpine Geology - a review. - *Wadia Commem. Vol., Min. Met. Inst. Ind.,* 171-194.

Kvale, A. 1957. Gefugestudien im Gotthardmassiv und den angrenzenden Gebieten. - *Schweiz. Min. Petr. Mit.,* **37**, 398-434.

Merh, S.S. 1966. On the Tectonic Pattern of the Central Kumaun Himalaya. *Pub. Cent. Adv. Stud. Geol.* (**1**)15.

Milnes, A. G. 1968: Strain Analysis of the Basement Nappes in the Simplon Region, Northern Italy. - 23rd Int. *Geol. Cong.,* **3**, 61-76.

Milnes, A.1973. Structural Reinterpretation of the Classic Simplon Tunnel Section of the Central Alps. - *Bull. Geol. Soc. Am.,* **84**, 269-274.

Naha, K., Ray S. 1970. Metamorphic history of the Jutogh Series in the Simla Klippe, Lower Himalaya. - *Contr. Mineral. and Petr.,* **20**, 117-184.

Plessmann, W. 1957. Zur Tektonik und Metamorphose der Bundnerschiefer am SW-Rande des Gotthardmassivs.- *Nachr. Akad. Wiss. Math.-Phys. Kl. IIa,* **1**, 18-31.

Plessmann, W. 1958. Tektonische Untersuchungen an Randteilen des Gotthard und Montblanc Massivs sowie an der Grenze Penninikum-Helvetikum. - *Nachr. Akad. Wiss., Math. Phys. Kl. IIa,* **7**, 153-188.

Saklani, P.S. 1971a. Structure and Tectonics of th Pratapnagar Area, Garhwal Himalaya. - *Himalayan Geology*, **1**, 79-91.

Saklani, P.S. 1971b. Occurrences of stromatolites in the Bhelunta Limestone formation of Pratapnagar Area, Garhwal Himalaya. - *Sci. and Cult.,* **37**(9), 438-439.

Saklani, P.S. 1972a. Lithostratigraphy and Structure of the Area between the Bhagirathi and Bhilangana Rivers, Garhwal Himalaya. *Himalayan Geology*, **2**, 342-355.

Saklani, P.S. 1972b. Geological Studies of the Area South of Mukhem, Garhwal Himalaya. - UnpublishedPh. D. Thesis, Delhi University, Delhi.

Saklani, P.S. 1973a. Metamorphic Petrology of the Area South of Mukhem, Garhwaḷ Himalaya. - *Recent Researches in Geology*, Hind. Publ. Corp., 82-106.

Saklani, P.S. 1973b. A review of the Himalayan orogenesis. - *Cas. min. geol.,* **18**, 193-198.

Saklani, P.S. 1974. Geologischer Aufbau des Mukhem Gebietes, Garhwal Himalaya, eine Zusammenfassung. - *Nachr. Deutsch. Geol. Ges.*, **10**, 20-21.

Schwan, W. 1965: Leitende Strukturen am Nordrand der Hohen Tauern. - *Verh. Geol. B.-A., Sonderheft G*, 214-245.

Thakur, V.C. 1973: Events in Alpine Deformation and Metamorphism in the Northern Pennine Zone and Southern Gotthard Massif Regions, Switzerland. - *Geol. Rdsch.*, **62**(2), 549-563.

Trumpy, R. 1960: Paleotectonic Evolution of the Central and Western Alps. - *Bull. Geol. Soc. Am.*, **71**, 843-908.

Valdiya, K.S. 1964: The tectonic design of the Himalaya, a survey of structures between Indus and Brahmaputra and their comparison with the Alpine structures. - 22nd. *Int. Geol. Cong.*, **11**, 283-315.

Wadia, D.N. 1932: The syntaxis of the North-West Himalaya, its rocks, tectonics and orogeny. - *Rec. Gol. Surv. Ind.*, **65**, 189-320, Calcutta.

Wenk E. (1969): Antithesis Alps - Himalaya. *Bull. Ind. Geol. Assoc.*, **1**, 33-36.

Wunderlich, H.G. (1958): Ablauf und Alterverhaltnisse der Tektonik-und Metamorphose Vorgange in Bundnerschiefern Nordtessins und Graubundens. - *Nachr. Akad. Wis., Math. Phys. Kl. IIa*, **7**, 115-151, Gottingen.

•□□□•

10

Thrust Tectonics of Garhwal

Many workers hold contrasting views about the Main Central Thrust (MCT). Valdiya (1978a, b) opined that the Vaikrita thrust is the real Main Central Thrust which also represents a metamorphic break between the high- and medium-grade metamorphics. The tectono-metamorphic contact between the low-grade gneisses and schists and the Pratapnagar quartzite was described by Fuchs and Sinha (1978), Valdiya (1981) and Pati and Rao (1983) as the Chail thrust. This contact was conceived by Gansser (1964) as the classical Main Central Thrust. In the opinion of Pati and Rao (1983) the contact between the high- and medium-grade metamorphics is not a tectonic one, but the contact between the medium- and low-grade gneisses and schists was designated by them as the Jutogh thrust and the conventional Main Central Thrust is their Chail thrust. Such inconsistent views about the Main Central Thrust (MCT) can be ascribed to its differential exposition and splitting caused by the multiple imbrication of the MCT I (Vaikrita), MCT II (Jutogh) and MCT III (Budhakedar, Saklani, 1993).

Saklani (1993) opines that the composite Main Central Thrust (MCT I, II and III) represents a major duplex structure. The roof part of the duplex is represented by the MCT I while the MCT III constitutes the floor part. This duplex is traversed by many overthrust faults and the lithologies occur in the form of tectonic horses (Fig. 10.1). A similar type of disposition of the Main Central Thrust was earlier established by Saklani and Bahuguna (1986) from the adjoining Bhagirathi - Bhilangna valleys of Garhwal Himalaya.

In the process of thrusting one body of rocks moves over the other along the thrust surface and infact, the plane behaves as a

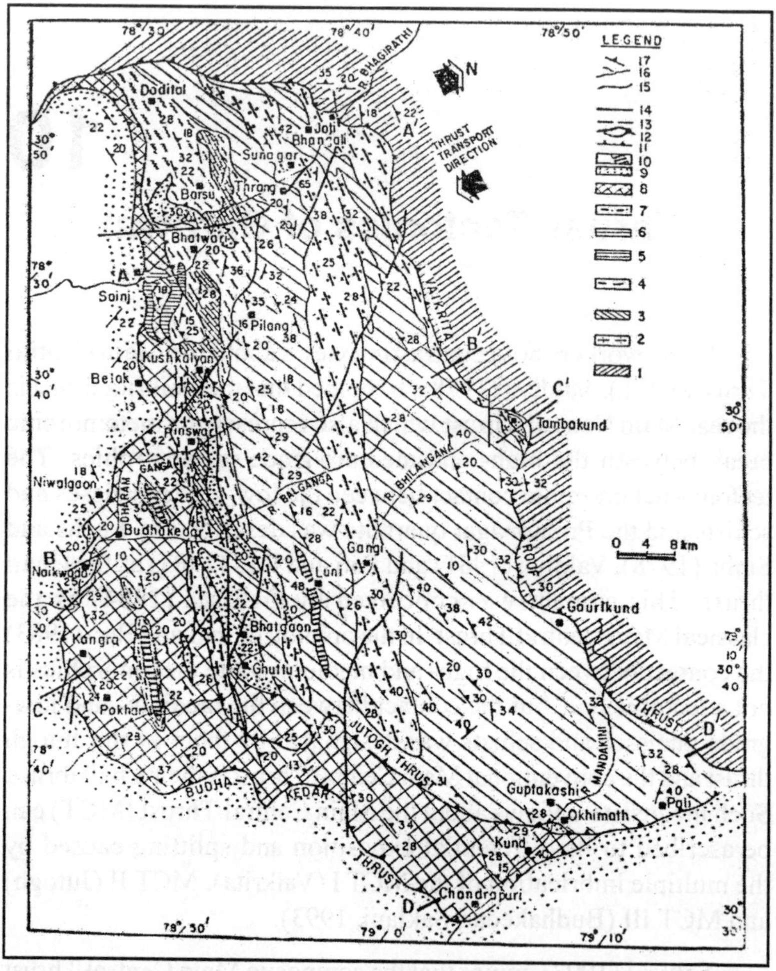

Figure 10.1: Geological Map of the Main Central Thrust and Associated imbricates. 1. Kyanite-sillimanite gneisses and schists, 2. Biotite schist, 3. Porphyroblastic gneiss, 4. Amphibolitic migmatite, 5. Gneissic migmatite, 6. Calcic migmatite, 7. Quartzofeldspathic schist, 8. Granite gneiss, 9. Quartzite, 10. Metabasics, 11. Thrusts, 12. Branch-line and branch-points, 13. Section lines, 14. Faults, 15. Lithological boundaries, 16. River, 17. Dip and Strike.

contractional fault. The rocks overlying a particular thrust surface belong to the hangingwall while the rocks below of it constitute the footwall. The thrust generally cuts up section in the transport direction rarely following a smooth plane (Dhalstrom, 1970).

During the thrusting the strained rock-bodies follow staircase trajectories from the point of view of the thrust movements consisting of ramps and flats and are chiefly governed by the topography of the footwall. In fact, a flat is parallel to bedding and is usually connected by steep ramps. These ramps can be classified on the basis of their disposition in the hangingwall and footwall. These can be further classified on their orientation with reference to the transport direction (Dhalstrom, 1969). As the deformation responsible for thrusting in the Himalayan region was of migratory type the thrusting in the Himalaya is of variable age.

According to Hossack (1983), the climbing up of thrust along the ramp in the direction of motion is an axiom. In such a type of deformation the younger thrusts are formed by the splitting of the footwall ramps of higher thrusts giving rise to the branch line which is formed at the leading edge of a lower/younger thrust. Because of the migratory nature of thrusting the branching or splitting of the older thrust takes place towards the foreland along its propagation direction. Trailing branch line follows the original shape of footwall ramps and the younger splays die out upward into the tip. Hence a part of branch line also represents the tip line.

The central part of Garhwal Himalaya is characterized by a branch line thrust at Barsu and Tambakund (Fig. 10.2) which is exposed on the surface as well as buried beneath the rocks. Because of extensive erosion the linking of the branch line thrust in the central part is, at times, conjectural and may be ascribed to hidden nature of the thrust. When such thrusts do not project up on the topographic surface they are termed as Blind thrusts (Thompson, 1981).

Migration of older thrust with formation of younger thrust is often controlled by forelimb of folding on a large scale. The migration is from forelimb to backlimb of the younger thrust. As such the forelimb in the hangingwall directly overlies the footwall or vice-versa. As far as the thrust transport direction is concerned, it is by nature towards the foreland. However, if the migrating younger thrusts acquire steep attitude due to vertical uplift, then the foreland directed bodies follow the piggy back mechanism in the footwall terrain and can reverse themselves to the hinterland

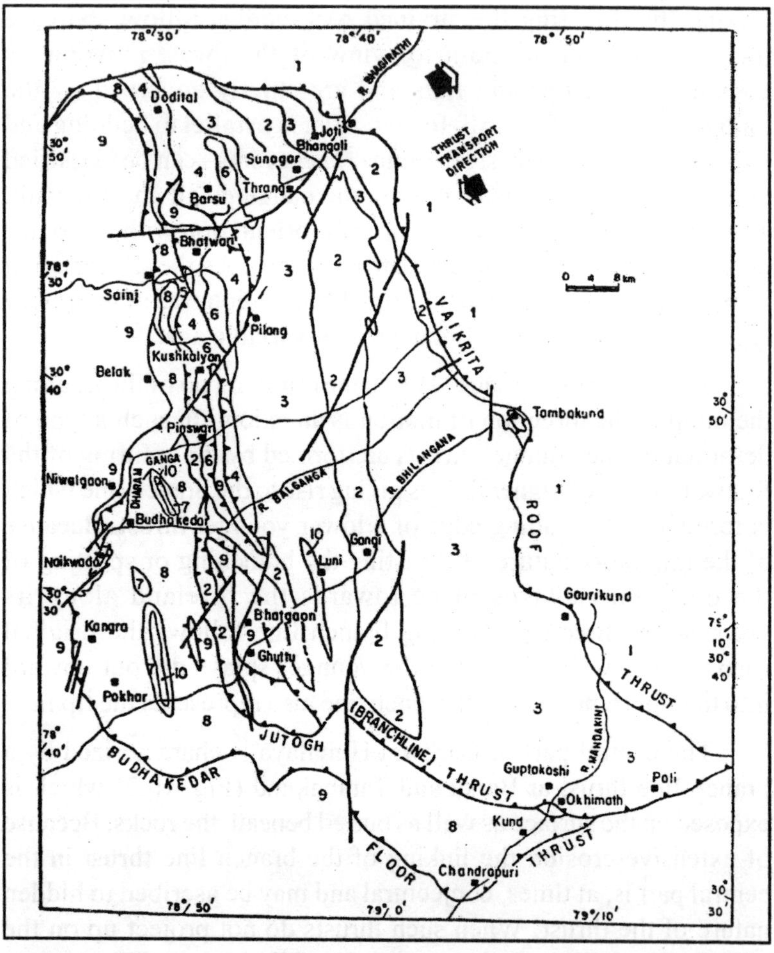

Figure 10.2: Structural map of the Main Central Thrust and associated imbricates. 1.Kyanite-sillimanite gneisses and schists, 2. Biotite schist, 3. Porphyroblastic gneiss, 4. Amphibolitic migmatite, 5. Gneissic migmatite, 6. Calcic migmatite, 7. Quartzofeldspathic schist,8.Granite gneiss, 9. Quartzite.

causing overstepping of the thrust slices i.e. the hangingwall zone of the thrust. The piggy back thrust sequence results from new thrusts being developed in the footwall of the older thrust (Butler, 1982). Apparently this occurs by progressive failure of footwall ramps and by abandonment of the footwall ramps (Cooper, 1981; Butler, 1982).

The continued failure of footwall ramps results in the formation of a schuppen structures (horses). When the series of asymptotic imbricate faults rejoin the earlier thrust at higher levels, together with a series of horses along with their bounding thrust-faults, then they are collectively grouped into a Duplex system. The higher thrust forms the roof thrust and last sharing with the roof is floor thrust. The duplex can have trailing and leading edges. When the roof and floor thrusts meet at the frontal part of the duplex it is the leading edge. A horse exhibits a single branch line at the intersection of two thrusts.

In the Garhwal Himalaya the Vaikrita (roof) Thrust and Budhakedar (floor) Thrust together with the Jutogh (branch line) Thrust constitute a major duplex which is characterized by a number of horses present in the Bhagirathi, Balganga, Bhilangana and Mandakini valley regions.

THRUSTS

(i) Vaikrita (Roof) Thrust

It extends from near Gauri Kund to Joti with a corrugated profile. It measures about 75 km in length along the strike forming the roof part of Central Crystalline Duplex System. The thrust is folded due to variation in size of the duplex along the strike and dip. The Tambakund window is just in the proximity of Vaikrita Thrust which is about 350 meters away from the window. The window has a length of about 50 meters. It thins down towards the NNW-SSE direction.

The flattening strains are indicated by the well developed chocolate tablet boudinage structures and folds in rocks exposed near Jhala village in the Bhagirathi valley. The development of penetrative fabric all along the Vaikrita (roof) Thrust in all the valleys, characterized by thick sequence of mylonites and blastomylonites indicates deformation belonging to deep structural levels. The ribbon structures and subgrain recrystallization are common in the schistose rocks of Joti formation, exposed near the Vaikrita (roof) Thrust. In the cross-sectional studies, it was found that the duplex developed due to collapse of the footwall ramp under the advancing root thrust. This thrust, thus, climbed up the

the rocks of Budhakedar formation. The slip within the roof thrust is variable because of the presence of transverse faults as a result of which the root thrust has been displaced (Fig. 10.2).

(ii) Jutogh (branch line) Thrust

It is exposed from Dodital in the Bhagirathi valley to Okhimath in the Mandakini valley (Fig.10.3). Lithologically, its splay is

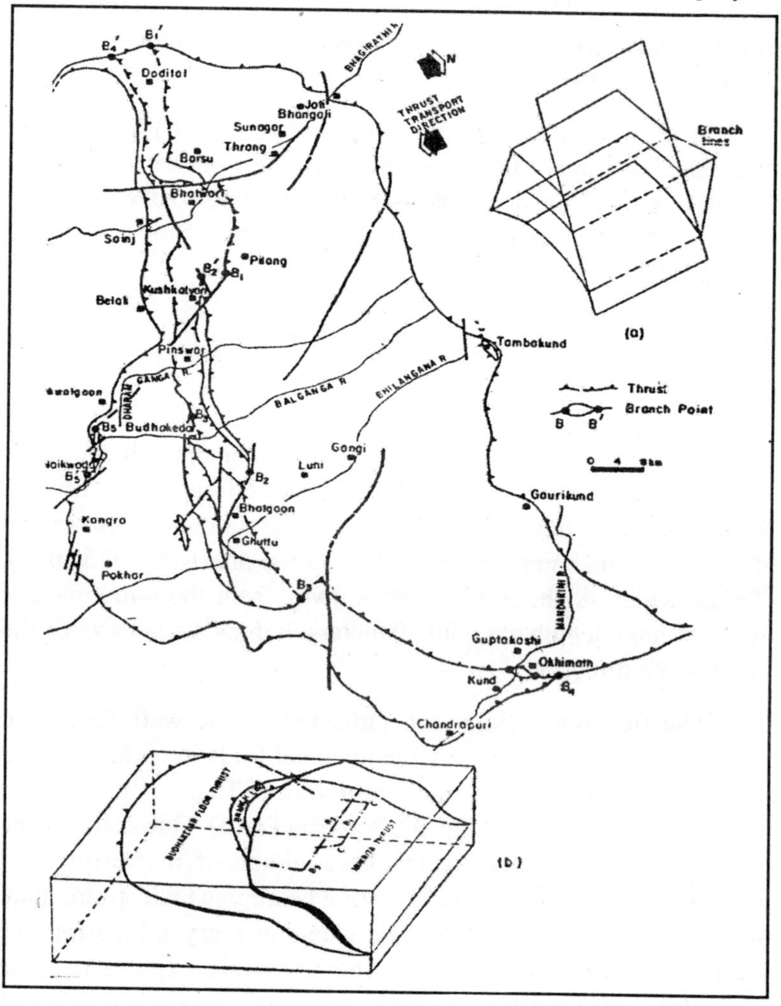

Figure 10.3: Branch-line map of the Main Central Thrust System. (a) Block diagram of branch-lines. (b) Schematic splaying of branch lines from the major Vaikrita Thrust.

mainly confined in the migmatites. When it meets the roof thrust then migmatites come in contact with the high grade rocks of Joti formation. Its incipient splay possibly started at the distal edges of the Barsu and Tambakund windows. The western edge of the Barsu window is joined with the roof (Vaikrita Thrust) while its continuation with the Tambakund and the eastern edge of the Tambakund is joined with the floor (Budhakedar thrust). The second splay with a major branch line (B_2-B_2') is exposed at about 13 km towards west of Bhatwari in the Bhagirathi valley and meets with the Vaikrita (roof) Thrust. It continues till Okhimath and meets the Budhakedar (floor) Thrust towards east (about 2 km) of Okhimath. This (B_2 -B_2') branch line is well defined westwards in the Balganga valley where it is exposed at the base of highly sheared biotite schist continuing upto the Bhilangana valley.

This thrust is faulted towards ENE (2.5 km) of Ghuttu. This branch line thrust is very well exposed on the surface and is characterized by extreme effects of shearing exemplified by the characteristic occurrence of mylonite and phyllonites.

Due to the presence of quartzitic windows the branch line thrust curves around the quartzitic windows, otherwise, it behaves like a single plane upto the floor thrust at Okhimath. In fact, there are three more branch line thrust-slices existent in the Garhwal region which may be ascribed to the migratory trend of deformation with associated fracturing of rocks as a result of which leading Branch line with advancing deformation front became trailing branch line thrust for the new leading branch line and vice versa.

The fourth branch line (B_4 -B_4 ') is exposed near Ghuttu and can be broadly linked up with the floor thrust in the Bhilangana valley at about 6 km east of Ghuttu and probably near Kushkalyan. This branch line thrust merges with the leading branch line thrust of the second stage.

In the advancing front of deformation the last branch line (B_5 -B_5 ') is seen in the Balganga valley near Naikwada and south of Niwalgaon and is characterized by quartzitic windows (Fig. 10.3).

(iii) Budhakeaar (floor) Thrust

It extends from Chandrapuri in the Mandakini valley in a zig-zag pattern to Sainj towards the northwest in Bhagirathi valley. It forms the floor part of the Central Crystalline Duplex System. The thrust is characterized by lenses and pockets of cataclasites around Pokhar and Kangra in Bhilangana valley and near Naikwada and Budhakedar in Balganga valley. The thrust is considered to be generated in the last stage of the thrusting movements (Main Central Thrust). It is characterized by the structures of higher level tectonics exemplified by quartz filled tension-gashes, en echelon veins and cataclasities associated with displaced lithologies. The presence of angular breccia zone was accompanied by high fluid pressure and the deformation was confined to narrow zones (5 meters in Naikwada area and about 1.5 meters in Chandrapuri area). Such an intense deformation in some areas (SE of Ghuttu) is wide-ranging. At times, the zone measures for about 50 to 70 meters in width as observed in Balganga, Bhilangana and Mandakini valleys. Fibre-crack fills and slickensides are found associated with the rocks of this thrust sheet, near Talyaban and Ghattugad (Balganga valley), Belak (Dharamganga valley) and Pokhar (Bhilanganga valley). It is generally found that the floor thrust formed at the synclinal part of Budhakedar formation.

CROSS-SECTIONAL STUDIES

Tectonic analysis of the thrust was carried out along four geological cross-sections and the current duplex length (L'), structural thickness (H'), stratigraphic thickness (t') and the area (A) was calculated. Initial length (Lo) of the duplex was determined by dividing the area of a cross section by stratigraphic thickness. The angles between the floor thrust and the central portion of subsiding faults was also measured. Current duplex length along the section line was measured in Bhagirathi valley section between the Vaikrita Thrust and Budhakedar Thrust. Since the stratigraphic units of the area are folded and no exact thickness of the formation is known so the stratigraphic thickness is inferred by taking the average of folded rock formations. Sections were restored to the undeformed state using area balancing method as proposed by

Dhalstrom (1969). A cross check was made by balancing individual horses. The sections can have some errors because the exact depth of decollement is unknown and could have been variable also. Further the stratigraphic thickness had to be taken on average basis in the absence of exact stratigraphic thickness. Development of sub-horses and refraction of the imbricate thrust faults into the bedding plane is another possible source of error. Due to these constraints sections were partially restored.

The various geological sections trending NE-SW have been balanced. It is presumed that all the rocks in the Garhwal region are in deformed state. By unfolding the deformed area in relation to the lithological thickness the original shape of the area can be estimated.

The deformation was because of shortening or contraction. It is, thus, clear that the folded, faulted and thrust lithologies on balancing can give a fair idea about the shortening ascribed to the deformation and based on the shortening distance, percentage of the shortening can be determined.

It may be remarked that overall shortening due to thrusting is generally less than that of the total shortening of the imbrication associated with the horses.

In preparation of the balanced cross section, current length (L') of the section from Vaikrita (roof) Thrust to the Budakedar (floor) Thrust in a horizontal direction was measured and the structural thickness of the lithologies occurring between the Vaikrita (roof) Thrust and Budhakedar (floor) Thrust along the topographic profile (H') was measured. These elements are associated with the lithologies and hence stratigraphic thickness is also very important for the balancing of the area. In most of the sections the average stratigraphic thickness (t') was apparent.

To determine the cross-sectional area between the Vaikrita (roof) and Budhakedar (floor) thrust the measurement of horizontal length (L') between the Vaikrita (roof) Thrust and the Budhakedar (floor) Thrust i.e. the central crystalline duplex and the structural thickness of the lithologies along the topographic profile, were helpful to determine the area (L'x H' = A).

The initial length (Lo) from the Vikrita (roof) Thrust i.e. the Central Crystalline Duplex System and Budhakedar Thrust can be determined by the following equation:

$$\frac{L' \times H'}{T'}$$

In this equation L' is current duplex length and H' is the structural thickness.

These parameters were very helpful to infer the shortening distance caused due to the thrusting and imbrication of rocks. The shortening distance was inferred by subtracting current horizontal length (L') measured between the Vaikrita (roof) Thrust and Budhakedar (floor) Thrust i.e. the Central Crystalline Duplex, from the initial length (Lo) of the same.

After the final correction of the deformed and partially restored sections the bulk shortening percentage from the geological cross-section was calculated by the following equation:

$ln\,[1+ (L'-L_o) / L_o] \times 100$ where *ln* is a natural long.

While dealing with the thrust models it was remarked that how the Vaikrita (roof) Thrust, Jutogh (branch line) Thrust, Budhakedar (floor) Thrust were splitted due to footwall collapse with the formation of ramps (Fig. 10.4). To observe this phenomenon with reference to ramp and flat trajectories of the different thrusts, restored sections were found useful to know about the thrust mechanism. In preparing the restored sections the structural thickness is not taken into consideration and only the length of the deformed stratigraphic unit is restored to its initial stage (Current duplex length L'). Had there been no deformation how the thrusting would have occurred along the ramps and flats can be visualized by the restored sections. From the balanced cross-sectional studies it was inferred that the Vaikrita (roof) Thrust and Jutogh (branch line) Thrust formed at an angle of 28° and 34° with respect to the Budhakedar Thrust and the angle was restored to show various trajectories of ramps and flats involved during the mentioned thrusting. The restored sections also gave an idea about the chronology of the slip transfer zones between the glide (thrust) horizons.

Bhagirathi Duplex

The Bhagirathi Duplex consists of upper and lower duplexes. The Vaikrita (roof) Thrust along with the Jutogh (branch line) Thrust defines the boundaries of upper duplex. The upper duplex is characterized by Barsu and Bhatwari tectonic horses while the lower duplex limits the Sainj horse. The upper limit of Barsu tectonic horse is defined by the Vaikrita Thrust while the lower limit of this is characterized by Barsu branch line Thrust. Bhatwari horse in the north is bound by Barsu branch line Thrust and in the south by Bhatwari branch line Thrust. The lower duplex (= Sainj tectonic horse) is demarcated in the north by the Jutogh branch line (Bhatwari) Thrust and in the south by the Budhakedar (floor) Thrust. Evaluation of shortening and contraction was carried out by measuring the current length (L') structural thickness (H') and stratigraphic thickness (t'). Tectonic elements of Bhagirathi Duplex together with the shortening and contraction are given in Table I.

The Bhagirathi Duplex is characterized by a current duplex length (L') of 22.7 kilometers and structural thickness (H') of 11.9 kilometers. The area of cross-section thus calculated with the help of L' x H' measures for about 270 square kilometers. Stratigraphic thickness (t') taken on an average basis measures 6.8 kilometers. The area of the cross-section when divided by the Stratigraphic

Table I: Tectonic Elements of Bhagirathi Duplex.

Current length L'	Structural thickness H'	Stratigraphic thickness t'	Cross-sectional area A	Initial length Lo	Shortening distance Lo-L'	Contraction ratio L'/Lo	Shortening percentage	ln [1+(L' - Lo)/Lo]×100
22.7 km	11.9 km	6.8 km	270 sq km	39.7 km	17.0 km	0.57	- 55.8%	Bhagirathi Duplex
21.85 km	10.4 km	5.7 km	227.2 sq km	39.86 km	18.0 km	0.54	-60%	Upper Duplex
14.5 km	7.4 km	3.7 km	107.3 sq km	29.0 km	14.5 km	0.5	-69%	Barsu horse
5.0 km	3 km	1.9 km	15 sq km	7.89 km	2.8 km	0.63	45%	Bhatwari horse
2.0 km	1 km	.9 km	2 sq km	2.2 km	0.2 km	0.90	-9.5%	Lower Duplex Sainj horse

thickness inferred an initial length (Lo) of 39.7 kilometers. On comparison between the current length (L') and initial length (Lo) of the Bhagirathi Duplex a difference of 17 kilometers was calculated. This value of 17 kilometers suggested that the Bhagirathi valley section witnessed a shortening distance of 17 kilometers. Based on the values of (L') and (Lo), Bhagirathi Duplex had a contraction of 0.57 and shortening was 55.8%.

As stated earlier, the Bhagirathi Duplex was divided into upper and lower duplexes and tectonic horses. The variation in shortening from upper duplex to lower duplex was attributed to the variable competence contrast and structural parameters of the rock types. The Barsu tectonic horse belonging to the upper duplex was shortened by -69% and was the maximum shortening along Bhagirathi Duplex. The minimum was -9.5% and was shown by lower duplex (Sainj horse).

In order to obtain the respective position and nature of different thrust faults along with the stratigraphic position of the rock types prior to deformation, restored sections were prepared (Fig. 10.4). In a restored section the area occupied by deformed lithologies was balanced by bringing them in their undeformed initial state. The technique (Dahlstrom, 1970; Boyer and Elliott, 1982) used to prepare the restored sections in the present work was mainly based on initial length (Lo) and stratigraphic thickness (t'). The restored section of the Bhagirathi Duplex (Fig. 10.4) reveals that the duplex prior to deformation was 39.7 kilometers in length. The total ramping up of Vaikrita Thrust from the sole thrust was 8.2 kilometers and it brought the rocks of Joti Formation to rest over the porphyroblastic gneiss of Bhatwari Formation and the Vaikrita Thrust then travelled as a flat for 12.1 kilometers at the top of porphyroblastic gneiss.

After this, the Vaikrita Thrust again ramped up for 3-5 kilometers and brought the high-grade Joti Formation to rest over another member of the Bhatwari Formation (i.e. the migmatites). In the second stage of thrusting, the Barsu branch line (Jutogh) Thrust ramped up 2.5 kilometers to merge with the Vaikrita Thrust and caused the formation of Barsu horse. Another branch line thrust

(Jutogh) i.e. the Bhatwari branch line thrust sheet, travelled for a distance of 17.1 kilometers as a flat from Barsu branch line in the southwest and ramped up for a distance of 3.2 kilometers to merge with the Vaikrita Thrust. This was responsible for the formation of Bhatwari horse. In the final phase the Budhakedar (floor) Thrust occurring at the base of granite gneiss (Budhakedar Formation) ramped up for 1.7 kilometers to merge with the Vaikrita Thrust and formed the Sainj horse (lower duplex).

Bhilangana Duplex

This major structure consists of upper and lower duplexes (Fig. 10.5). The upperduplex is defined by the Vaikrita (roof) and Jutogh branch line (floor) Thrust from north to south.

In the upper duplex, three tectonic horses namely, Tambakund horse, Luni horse and Northeastern Ghuttu horse are present. The lower duplex is also represented by the southeastern Ghuttu horse. The northern most tectonic horse (Tambakund) is bound in the north by Vaikrita Thrust and in the south by Tambakund branch line Thrust. The Luni horse is bordered in the north by Tambakund branch line thrust and in the south by Ghuttu branch line thrust. The boundaries of Northeastern Ghuttu horse are delineated by Ghuttu branch line thrust in the north and Jutogh Thrust in the south. The lower duplex is marked by Jutogh thrust in the north and Budhakedar thrust in the south. The Bhilangana Duplex has current length (L') of 41.7 kilometers and structural thickness (H') of 23.2 kilometers. The cross-sectional area (A') of the Bhilangana Duplex measures 963.2 square kilometers. The initial length (Lo) of the duplex is 166 kilometers. The difference of 124 kilometers between the Lo-L' infers that the duplex witnessed contraction of 0.25 with -138% shortening. The Tambakund horse underwent -160.9% of shortening and contraction of 0.2. The Southeastern Ghuttu horse showed 150% shortening and 0.22 contraction (Table II).

It was inferred from the restored section (Fig. 10.5) that at a distance of 166 kilometers from southwest, the Vaikrita Thrust ramped up for about 7.5 kilometers and brought the rocks of Joti Formation over the rocks of Bhatwari Formation. The Vaikrita

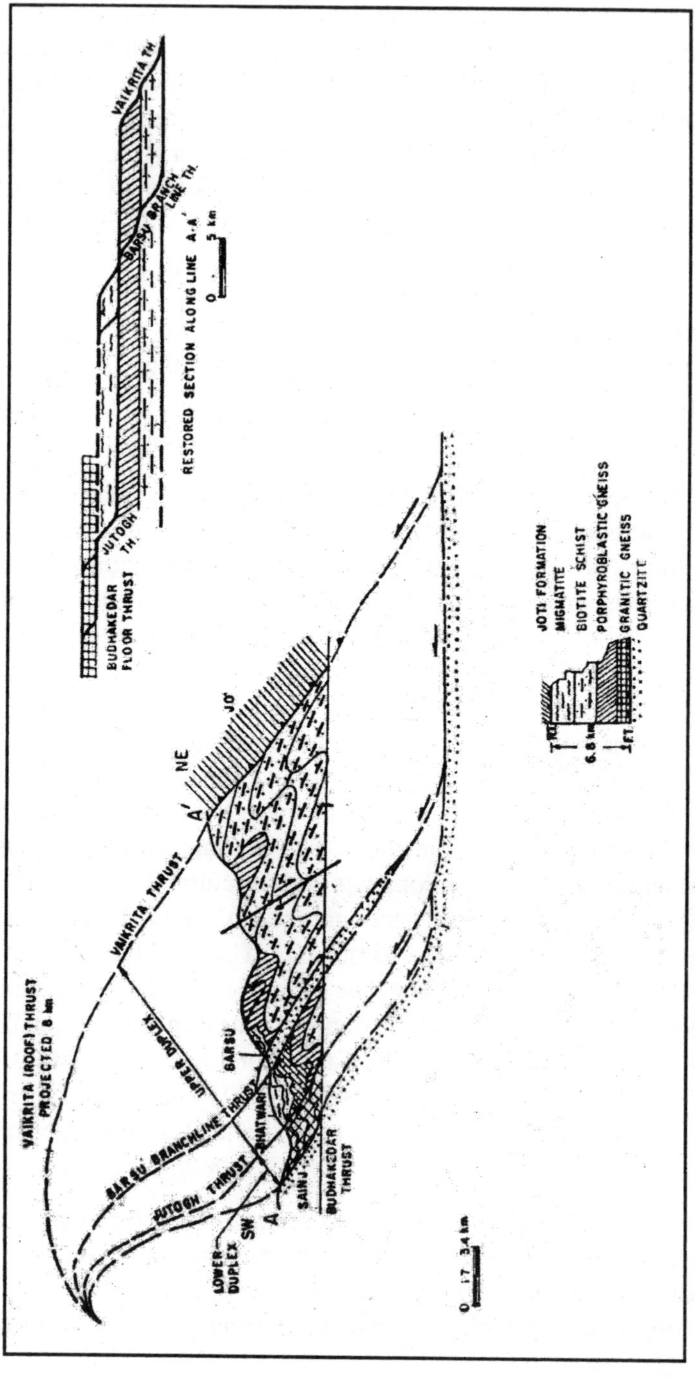

Figure 10.4: Geological Cross-Section and Restored Section along line A-A′ Kyanite-sillimanite gneisses and schists, 2. Biotite schist, 3.Porphyroblastic gneiss 4. Amphibolitic migmaties, 5. Calcic migmatite, 6. Granite gneiss, 7. Quartzite.

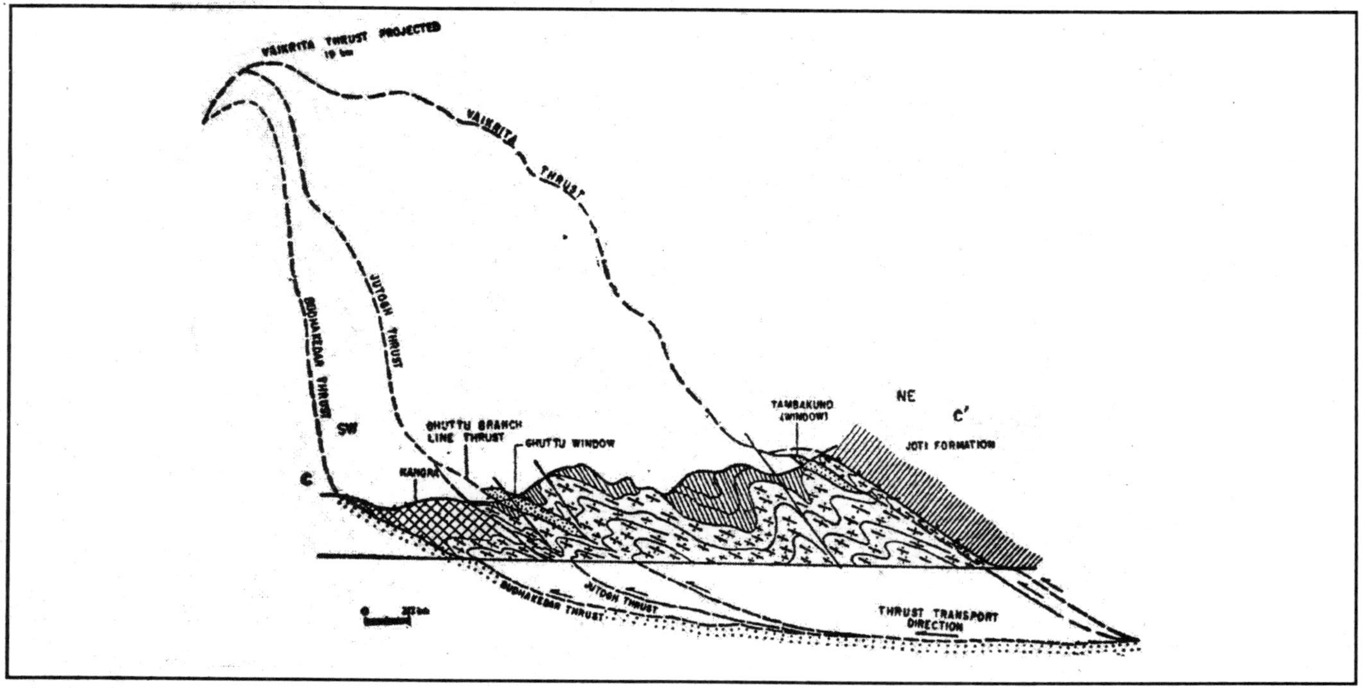

Figure 10.5: Geological Cross-Section along line C-C'. Kyanite-sillimanite gneisses and schists, 2. Biotite schist, 3. Porphyroblastic gneiss 4. Granite gneiss, 5. Quartzite.

Table II: Tectonic Elements of Bhagirathi Duplex.

Current length L'	Structural thickness H'	Stratigraphic thickness t'	Cross sectional Area A	Initial length Lo	Contraction ratio L'/Lo	Shortning distance Lo - L'	Shortening percentage	ln [1 + (L' - Lo)/Lo] × 100
1.15 km	1 km	0.2 km	1.15 sq km	5.75 km	0.2	4.25 km	- 160.9%	Tambakund horse
28.4 km	15.1 km	3.75 km	428.84 sq km	114 km	0.24	85.9 km	- 138.9%	Luni horse Upper Duplex
4.35 km	1.8 km	0.4 km	7.83 sq km	19.5 km	0.22	15.22 km	- 150%	Northeastern Ghuttu horse
10.8 km	5 km	1.7 km	54 sq km	31.76 km	0.34	20.96 km	- 107.8%	Southeastern Ghuttu horse Lower Duplex
41.7 km	23.1 km	5.8 km	963.2 km	166 km	0.25	124 km	-138%	Bhilangana Duplex

Thrust then travelled for a distance of 137 kilometers as a flat on the top of porphyroblastic gneiss of Bhatwari Formation. The Tambakund (branch line) Thrust, however, ramped up for a distance of 7 kilometers towards southwest. The Ghuttu branch line thrust ramped up for about 8.9 kilometers to merge with the Vaikrita Thrust located at the top of porphyroblastic gneiss of Bhatwari Formation. The floor thrust travelled for a distance of 21 kilometers and then ramped up by 6.7 kilometers to meet with the Vaikrita Thrust. In the last stage, the slip was transferred from the base of porphyroblastic gneiss of Bhatwari Formation to the base of granite gneiss of Budhakedar Formation. During this the floor thrust travelled for 0.7 kilometers at the base of granite gneiss and ramped up for 2.1 km to form lower duplex.

Mandakini Duplex

The upper duplex (Fig. 10.6) is marked by the Vaikrita (roof) Thrust and (floor) by the Jutogh (branch line) Thrust. The lower duplex is bound in the north by Jutogh Thrust and in the south by the Budhakedar (floor) Thrust.

The Mandakini Duplex has a current length (L') of 16-25 kilometers and structural thickness (H') of 13 kilometers. The cross-sectional area of the duplex measures about 211 square kilometers. The shortening and contraction (Table III) witnessed by the Duplex measured -69% and.0.50 with initial length (Lo) of 32.5 kilometers. The shortening distance [or the difference of current length (L') and initial length (Lo)] was 16.2 kilometers.

Restored section of the Mandakani Duplex (Fig. 10.7) inferred that the Vaikrita Thrust ramped up by 4.2 kilometers from the sole

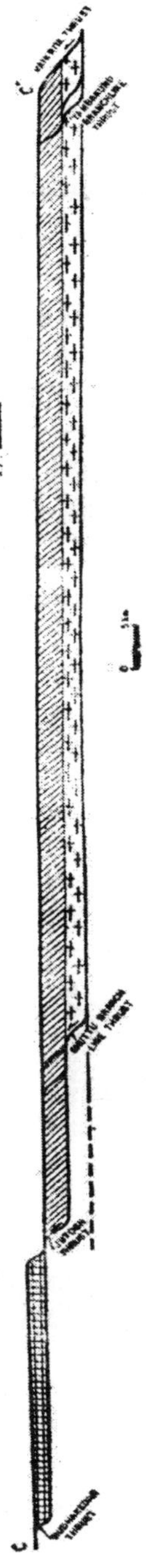

Figure 10.5: Restored Cross-Section along line C-C′ in Figure 10.5.

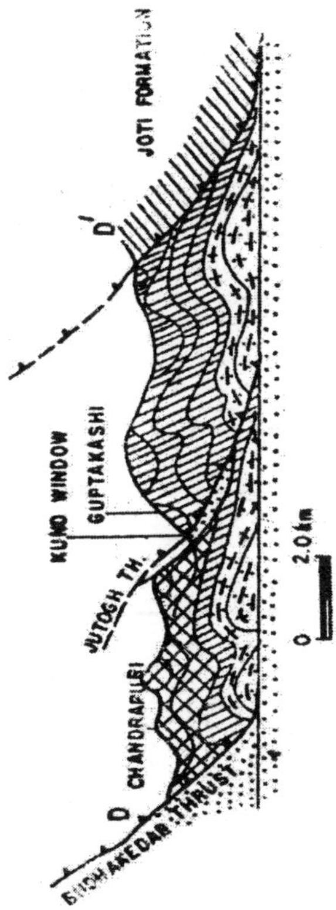

Figure 10.6: Cross-Section along line D-D′.

thrust and brought the rocks of Joti Formation to rest over the porphyroblastic gneiss of Bhatwari Formation. The Vaikrita Thrust travelled as a flat on the top of porphyroblastic gneiss for a distance of 7.8 kilometers before it changed its level from the top of porphyrobbstic gneiss to granite gneiss of Budhakedar Formation. During the second stage, the Jutogh Thrust ramped up by 4.2 kilometers to form upper duplex. The sole thrust changed its disposition from the porphyroblastic gneiss to the granite gneiss. This thrust occurring at the base of granite gneiss for 16 kilometers cut through the rocks for 57 kilometers to form the lower duplex structure. Based on the analysis of all the three duplexes along geological cross sections A-A', B-B' and D-D' it was inferred that the average shortening witnessed by the duplexes was 87.6°%.

Table III: Shortening distance, percentage and contraction of the Bhagirathi, Bhilangana and Mandakini Valleys.

Units	Bhagirathi valley			Bhilangana valley			Mandakini valley		
	Shortg. Distance	Contraction ratio	Shortg. %	Shortg. Distance	Contraction Ratio	Shortg. %	Shortg. Distance	Contraction Ratio	Shortg. %
Jutogh (upper) nappe	0.2 km	0.90	-9.5	85.9	0.24	-139	4.2	0.50	-65
Budhakedar nappe (lower)	18 km	0.54	-60	20.96	0.34	-107	12	0.50	-69
Bhagirathi Duplex	17 km	0.57	-55	124	0.25	-136	16.2	0.50	-69

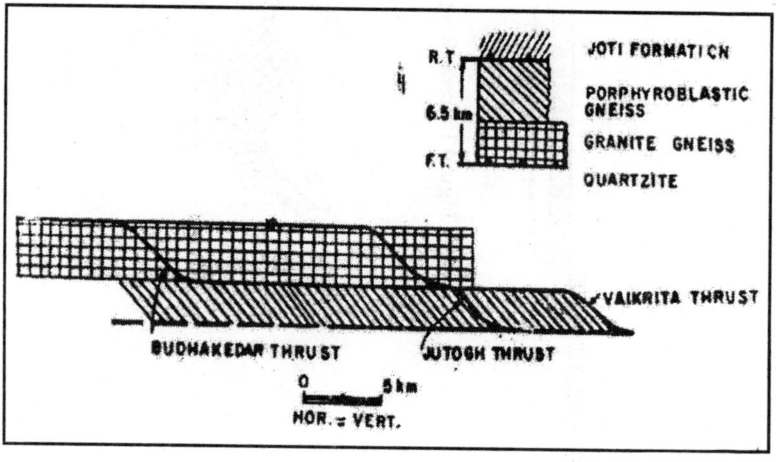

Figure 10.7: Restored Cross Section along line D-D' in Figure 10.6.

YAMUNA VALLEY REGION

The Yamunotri Valley region of Garhwal Himalaya also consists of the three planes of the Main Central Thrust exposed near Yamnotri, Jankichatti-Phulchatti and Wazri. The thrust-bound Central Crystallines and metamorphics constitute the hangingwall imbricates resting over the quartzites of the Pratapnagar Group. The quartzites always occur at the footwall side of the Main Central Thrust (Figs. 10.8, 10.9).

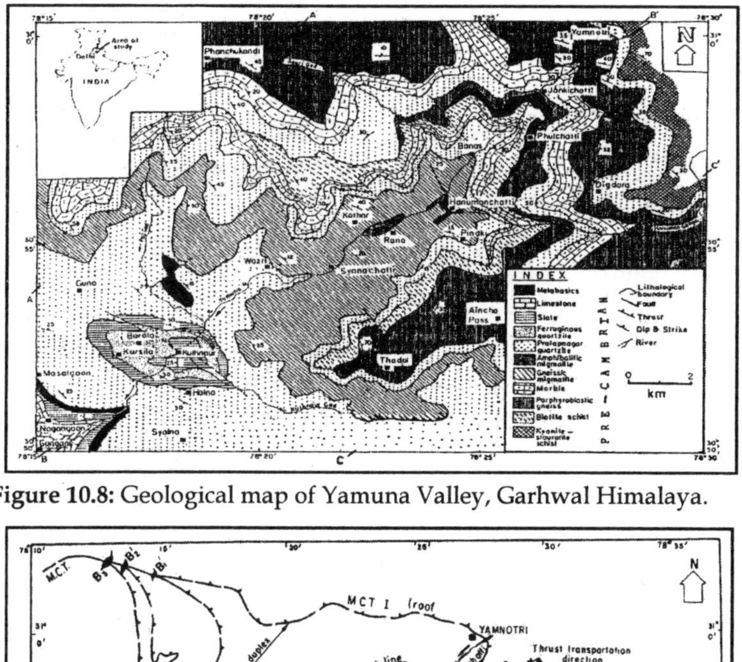

Figure 10.8: Geological map of Yamuna Valley, Garhwal Himalaya.

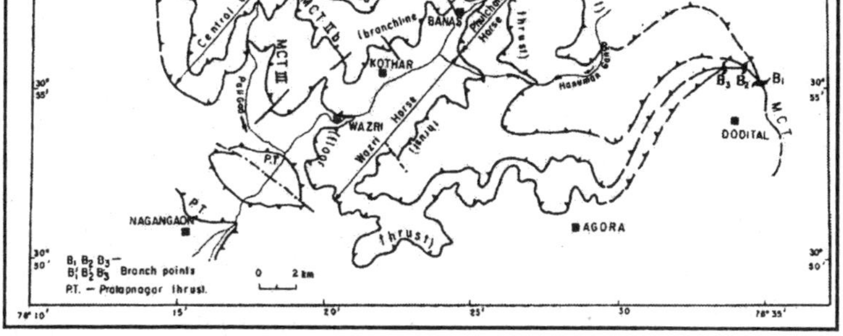

Figure 10.9: Disposition of Central Crystalline duplex with branch-line thrust of Yamuna Valley region.

Three representative sections were selected for tectonic analysis and based on current duplex length (L'), structural thickness (H'), stratigraphic thickness (t') and cross-sectional area (A) and the shortening of the duplex in the Yamuna Valley region was inferred.

GENERAL GEOLOGY

Predominantly the Garhwal region is constituted by the Central Crystalline Group of gneisses, schists with marble bands and migmatized granite gneiss representing high, medium and low grades of metamorphics (Pachauri, 1972, Agarwal and Kumar 1973). The high-grade metamorphism consist of kyanite - staurolite schist (Yamnotri Formation). It has been designated as the Vaikrita group by Jangpangi (1976) and Valdiya (1978, 1980). The Biph Formation is of medium grade and is constituted by porphyroblastic gneiss, biotite schist with marble bands. Valdiya (1978, 1980) described it as the Munsiari Formation in the Kumaun Himalaya. Pati and Rao (1983) called it the Jutogh Formation in Garhwal Himalaya. The low-grade migmatized granite gneiss is included into the Syanachatti Formation. Narayan Das et al. (1981) remarked that the granite gneiss (low-grade) in the adjoining Bhilangna valley ranges in age from 1178 ± 44 to 1862 ±37 m. y. Infact, the entire lithostratigraphic column belongs to the Pre-Cambrian in age.

In the field, the high, medium- and low-grade rocks frequently show their tectonic contacts with the Pratapnagar quartzite (Fig. 10.8). The Pratapnagar/Berinag quartzite has been assigned an age of 1230 ± 100 to 1232 ± 120 m. y. by Narayan Das et al. (1981). It may be remarked that the MCT II was much affected by imbrication and as such it can be subdivided into two tectonic planes MCT IIa and MCT IIb. Infact, it is a branch line thrust (Fig. 10.9). The MCT IIa brings the porphyroblastic gneiss and marble over the Pratapnagar quartzite while the porphyroblastic gneiss, biotite schist and marble are thrust over the Pratapnagar quartzite along the MCT IIb. The argillo-calcareous Kotga-Banali group (Kuthnaur Fm.) occurs in the form of a window bound by the Pratapnagar Thrust. The tectonic setting of the area is given in Table IV.

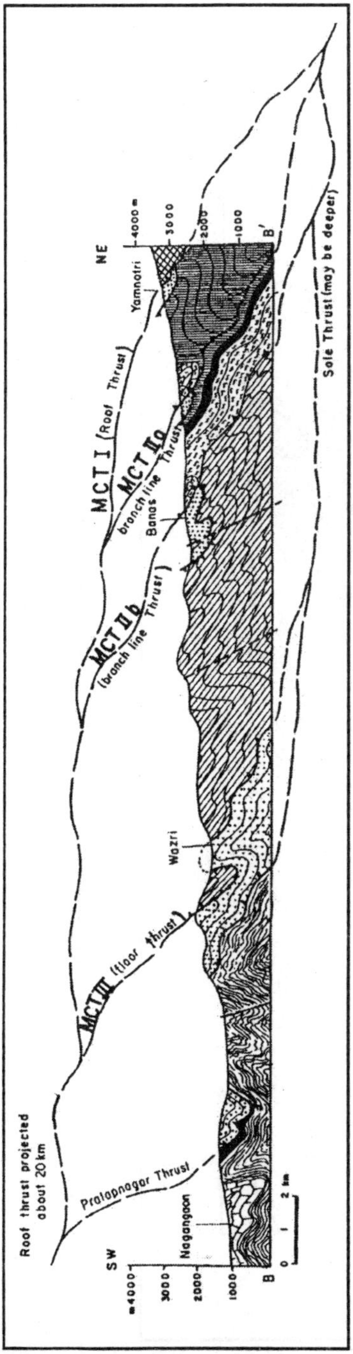

Figure 10.10: Geological cross sections. – A; Along line A-A′.

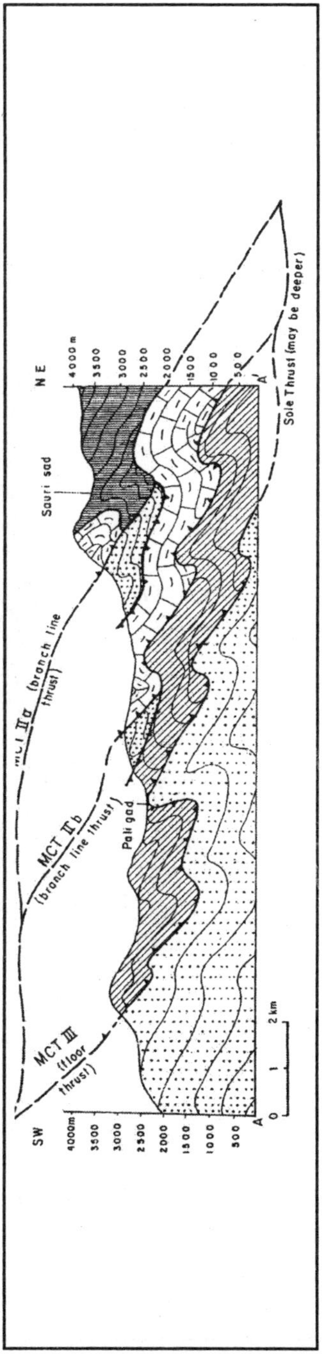

Figure 10.11: B. Along line B-B′.

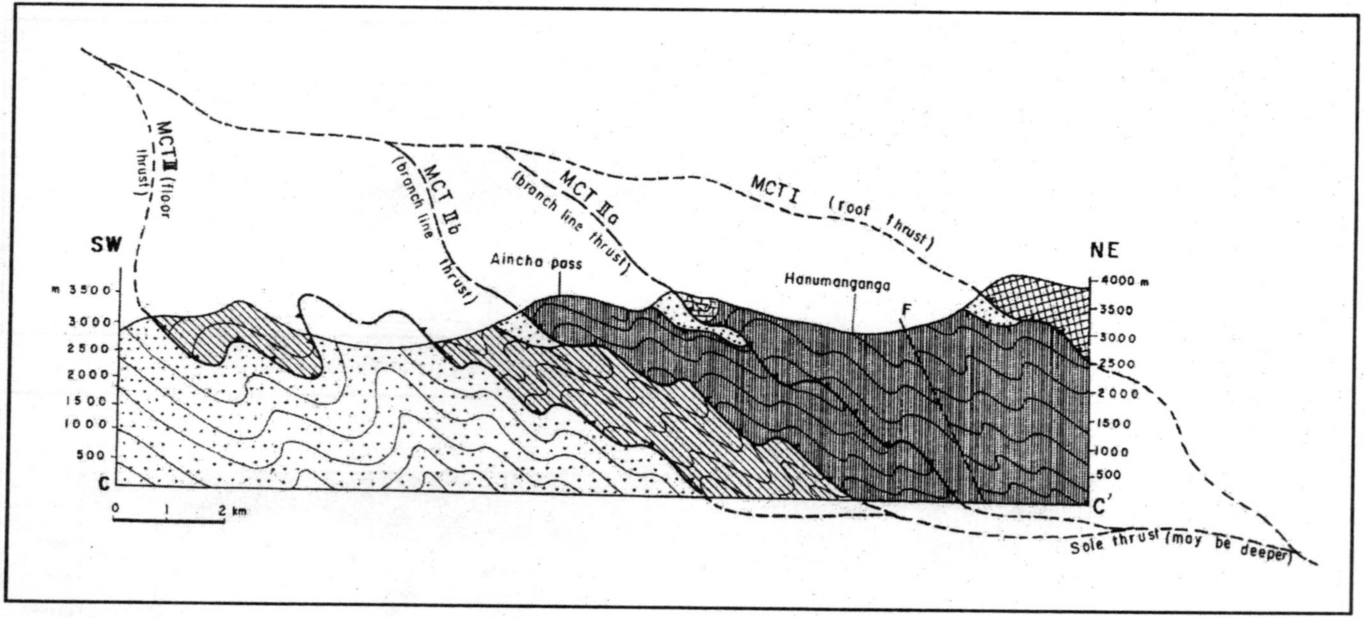

Figure 10.12: Geological cross sections. - Along line C-C.

Table IV: Tectonic setting of the area.

Thrust	Lithology
	Kyanite - Staurolite Schist
----- MCT I (Vaikrita) -----	
	Porphyroblastic gneiss with marble
	----- MCT IIa -----
----- MCT II (Jutogh)	Porphyroblastic gneiss and Biotite schist with marble
	----- MCT IIb -----
	Migmatized granite gneisses
----- MCT III (Chail?) -----	
	Mylonitized and massive fractured quartzite
----- Pratapnagar Thrust -----	
	Limestone, slate and ferruginous quartzite

Table V: Tectonic elements of Central Crystalline duplex (A-A′ section).

Structural Units	Current length L' (km)	Structural thickness H' (km)	Stratigraphic thickness t' (km)	Cross-sectional area A (Sq km)	Initial length L_o (km)	Shortening distance L_o-L' (km)	Shortening* percentage	Contraction ratio L'/L_o
Yamnotri - Nagangaon Section	19.3	10.2	4.9	197	40	20.7 (about 21)	-73%	0.48
Jankichatti horse	3.5	2	1.3	7	5.5	2	-43%	0.65
Phulchatti horse	3	1.6	1	4.8 (about 5)	4.8 (about 5)	1.8 (about 2)	-45%	0.64
Wazri horse	11.5	6.5	2.6	75	29	17	-92%	0.40

* $\ln(1+[L'-L_o]/L_o) \times 100$

CRYSTALLINE DUPLEX OF YAMUNA VALLEY

The tectonic analysis of thrust slices was carried out with the help of geological cross sections viz. (i) Phanchukandi-Guna (A-A'), (ii) Yamnotri - Nagangaon (B-B') and (iii) Digdara - Thadol (C-C'). The terminology of the thrust system as proposed by Boyer and Elliott (1982) and Butler (1982) was adopted for the analysis. The horizontal distance between the MCT I (roof) and MCT. III (floor) along the geological profiles is designated as the current duplex length (L'). The thickness of the deformed lithounits between these two thrust planes, perpendicular to L' in a vertical plane, is denoted as structural thickness (H').

The stratigraphic thickness (t') of individual stratum is variable due to folding. All the rocks are unfossiliferous, too. Therefore, the determination of the original thickness was problematic. Based on the rock-dips (structural contours) and the effect of folding the average stratigraphical thickness is inferred. (Johnson, 1986), Bahuguna and Saklani (1988) and Le fort (1989) also used this method to measure stratigraphical thickness in deformed rocks of the Garhwal and Nepal Himalayan regions.

With the help of these three parameters the cross-sectional area (A) = L' x H' and the initial length (L_0) - A/t' of the Central Crystalline duplex were determined. The bulk shortening percentage of Central Crystallines and metamorphics was determined by the equation $ln\ (1 + [L' - L_0]/L_0) \times 100$ (where *ln* is a natural log).

Phanchukandi-Guna section (A-A') (Table IV)

In the Phanchukandi-Guna section, medium-grade porphyroblastic gneiss and marble overlie the Pratapnagar quartzite along the MCT IIa and MCT IIb (branch line thrust) and the marble unit occurs in the form of a medium-grade Phulchatti horse. Similarly, the low-grade migmatized granite gneiss tectonically rests over the Pratapnagar quartzite along the MCT III and the low-grade Wazri horse is constituted by the granite gneiss (Fig. 10.10, 10.11). It may be remarked that along the Phanchukandi-Guna section only the medium-grade Phulchatti horse is present. The MCT IIa represents the roof while the MCT IIb is the floor part of this thrust (= Jutogh Thrust). It was subsequently followed by the low-grade Wazri horse.

The current length L' of the duplex (Fig. 10.10, 10.11) measures for about 13.5 km. The structural thickness (H') is for about 5.25 km with an average stratigraphic thickness (t') of 3.35 km. With the help of these three parameters the cross-sectional area (A) is determined for 71 sq km and the original length (L_0) for 21 km. The shortening distance (L_0-L') as calculated for the medium-grade Phulchatti and low-grade Wazri horses was about 7.5 km (- 45 % shortening).

(i) Medium-grade Phulchatti horse

The current length (L') of the medium-grade Phulchatti horse is of 3.5 km. The deformed thickness (H') of the lithologies (about

Table VI: Tectonic elements of Central Crystalline duplex (B-B′ section).

Structural Units	Current length L' (km)	Structural thickness H' (km)	Stratigraphic thickness t' (km)	Cross-sectional area A (Sq km)	Initial length L_o (km)	Shortening distance L_o-L' (km)	Shortening* percentage	Contraction ratio L'/L_o
Yamnotri - Nagangaon Section	19.3	10.2	4.9	197	40	20.7 (about 21)	-73%	0.48
Jankichatti horse	3.5	2	1.3	7	5.5	2	-43%	0.65
Phulchatti horse	3	1.6	1	4.8 (about 5)	4.8 (about 5)	1.8 (about 2)	-45%	0.64
Wazri horse	11.5	6.5	2.6	75	29	17	-92%	0.40

* ln (1+[L'-L_o]/L_o) × 100

2.25 km) is three times more than the average stratigraphic thickness (t': 0.75 km). The area of cross section (A) measures for 8 sq km. The initial length (L_0) measures for 10.5 km which is suggestive of the fact that the medium-grade Phulchatti horse was shortened by 7 km (i.e. -110% of shortening) with a contraction ratio of 0.33.

(ii) Low-grade Wazri horse

The low-grade Wazri horse has a current length (L') of 6 km. The structural thickness (H') measures for 3 km and the stratigraphic thickness (t') is for 2.6 km. The cross-sectional area of this horse measures for about 18 sq km with an initial length of 7 km. The low-grade Wazri horse was shortened by 1 km and witnessed shortening of- 14% with a contraction ratio of 0.86.

The comparative data of the Phulchatti and Wazri horses pertaining to shortening parameters are given in Table II.

***(iii) Restored section** (Fig. 10.13)*

The restored section along the Phanchukandi-Guna section (A-A1) reveals the initial length (L_0) measures for 21 km (Fig. 10.13). It depicts that the MCT IIb was flat for 10.6 km (about 11 km) while the MCT IIb ramped up for about 1 km and merged with the already existing MCT IIa. The movement of the MCT III (floor thrust) was horizontal along a flat for about 7 km with respect to the previously existing MCT IIb and at this stage it ramped up for about 4 km.

Yamnotri-Nagangaon section (B-B')

The major Central Crystalline duplex along the Yamnotri-Nagangaon section consists of the medium-grade Jankichatti-, Phulchatti- and low-grade Wazri-horses.

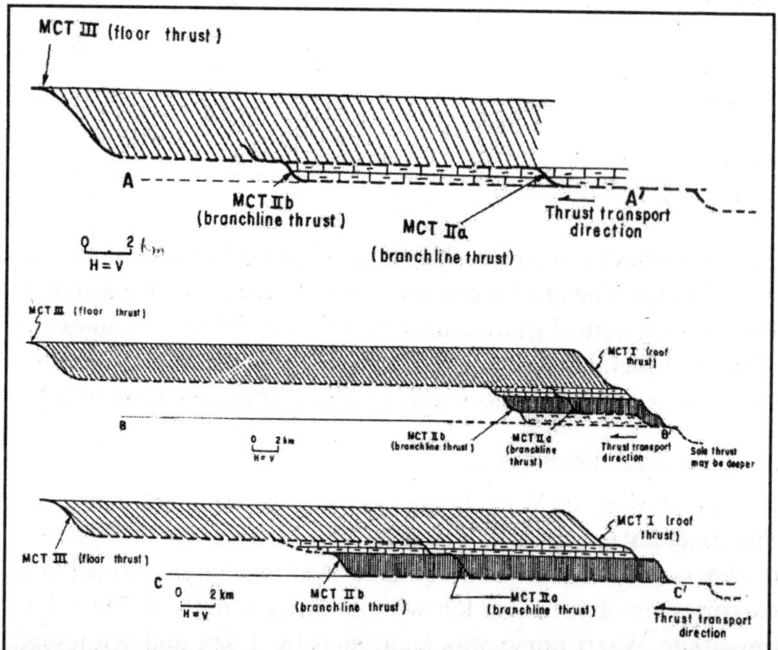

Figure 10.13: Restored sections based on initial length (L0). – A; Along line A-A' : B: Along line B-B'; C: Along line C-C'.

It has a current length (L') of 19.3 km (i.e. 19 km), the structural thickness (H') measures for about 10 km with an average stratigraphic thickness (t') of about 4.9 km (i.e. 5 km). The area measures for about 197 sq km (Tab. III). The initial length (L_0) of the lithologies is for about 40 km. The lithologies were shortened by 21 km (i.e., - 73 % of shortening with a contraction ratio of 0.48).

(ia) Medium-grade Jankichatti horse

Its current length (L') is 3.5 km and the structural thickness is 2 km. The average stratigraphic thickness is 1.3 km. The cross-sectional area (A) measures for about 7 sq km with an initial length

(L_0) for about 5.5 km. The shortening distance of the horse is inferred for about 2 km (- 43 % of shortening) with a contraction ratio of 0.65.

(ib) Medium-grade Phulchatti horse

The L' along the medium-grade Phulchatti horse is 3 km. The structural and stratigraphic thickness measure as 1.6 km and 1 km respectively. The area of the horse along this section is about 5 sq km with L_0 of 5 km. The shortening distance inferred is about 2 km with - 45% of shortening (contraction ratio-0.64).

(ii) Low-grade Wazri horse

The current length (L') of the horse is about 11.5 km and the structural thickness (H') is 6.5 km. The average stratigraphic thickness is 2.6 km. The cross-sectional area (A) measures for about 75 sq km with an initial length of about 29 km. The Wazri horse was shortened for about 17 km (- 92% of shortening) with a contraction ratio of 0.4.

(iii) Restored section

The Yamnotri-Nagangaon section (Fig. 10.13) of the Central Crystalline duplex reveals that the MCT I (roof thrust) ramped up by 8.5 km from the MCT III (floor thrust) and during the process, intersected the medium- and low-grade metamorphics. The lithologies along the MCT I travelled horizontally along a flat for a distance of about 40 km over the low-grade migmatized granite gneiss. Due to the migration of stresses the medium-grade porphyroblastic gneiss and schists along the MCT IIb ramped up for about 5 km, which along the MCT IIa travelled over a flat surface for about 5 km and later ramped up about 4 km to lie over the low-grade migmatized granite gneiss along the MCT IIb which was horizontal for about 29 km and then vertically ramped for about.

Digdara-Thadol section (C-C')

The duplex of crystallines and metamorphics is constituted by medium-grade Jankichatti-, Phulchatti- and the low-grade Wazri horses.

Along the Digdara-Thadol section (Fig. 10.12), the current duplex length (L') measures for about 16.75 km. The structural thickness (H') and the stratigraphic thickness (t') measure for about 7.1 km and 3.8 km respectively. The cross-sectional area (A) measures for 119 sq km. The calculated initial length (L_0) of the lithologies is for 31 km. The duplex was shortened for 14 km and witnessed-63% shortening with a contraction ratio of 0.53.

(i) Medium-grade horses

The medium-grade metamorphics of the Jankichatti horse shows current length (L') for 5.5 km and their initial length is for about 12 km. This medium-grade Jankichatti horse, due to deformation, shortened for 6.5 km with-79% of shortening. In contrast to this, the Phulchatti horse measures for about 2.75 km and is about half the size of the Jankichatti horse. It witnessed -93 % of shortening.

(ii) Low-grade Wazri horse

This horse has 6.6 km of current length and 12 km of the initial length of the low-grade rocks. It witnessed -61% of shortening with a contraction of 0.54 (Table VI, VII).

(iii) Restored section

The total horizontal distance from the MCT III to MCT I is for about 31 km. Initially the MCT I ramped up for about 7.5 km (Fig. 10.13). Subsequently, the sheet of the MCT IIa horizontally travelled along a flat for about 12 km and then ramped up for about 3 km. Some of the stresses migrated horizontally at the base of the porphyroblastic gneiss for about 7 km (flat). The MCT IIb ramped for about 2 km. The stresses further migrated southward along the flat (12 km). The subsequent ramping of the MCT IIb for about 3 km made this thrust to merge with the MCT III.

DISCUSSION

As described in the preceding text, the imbricated zone of the MCT II is characterized by the Jankichatti (MCT IIa) and Phulchatti (MCT IIb) horses.

Table VII: Tectonic elements of Central Crystalline duplex (C-C′ section).

Structural Units	Current length L' (km)	Structural thickness H' (km)	Stratigraphic thickness t' (km)	Cross-sectional area A (Sq km)	Initial length L_0 (km)	Shortening distance L_0-L' (km)	Shortening* percentage	Contraction ratio L'/L_0
Digdara-Thadol section	16.75	7.1	3.8	119	31	14	−63%	0.53
Jankichatti horse	5.5	2.75	1.25	15	12	6.5	−79%	0.45
Phulchatti horse	2.75	1.9	0.75	5	6.9 (about 7)	4	−93%	0.39
Wazri horse	6.6	3.3	1.8	22	12	5.5	−61%	0.54

* $\ln(1+[L'-L_o]/L_o) \times 100$

The Jankichatti horse of the MCT IIa along different sections shows variable current length (L') ranging from 3.5 to 5.5 km. Prior to the deformation, the lithologies of the horse measured for about 5.5 to 12 km. During thrusting these were shortened from -43 to -79% (shortening distance from 2 to 6.5 km).

With regard to the Phulchatti (MCT IIb) horse along different sections, the current length (L') of the rocks varies from 2.75 to 3.5 km which initially measured from 5 to 10.5 km. It witnessed-45 to-110% of shortening (shortening distance of 2 to 7 km).

The Wazri horse belonging to the MCT IIa along different sections shows a current length (L') varying from 6 to 11.5 km. Prior to deformation the lithologies of the Wazri horse along the MCT III measured from 7 to 29 km and were shortened -14 to -92% with a shortening distance of 1 to 17 km.

In general, the medium-grade Jankichatti and Phulchatti horses during the formation of the MCT II underwent for about -70% of shortening with shortening distance of about 8.5 km while the low-grade Wazri horse exposed along the MCT III was shortened by -55 % (shortening distance by 7.9 km or about 8 km).

A comparative study of the restored sections indicates that the MCT I during the thrusting ramped for about 8 km and then this sheet (= Vaikrita) of high-grade gneisses and schists was horizontally translated for about 31 to 40 km to merge with the MCT III. The kyanite-staurolite schist was the main rock which was affected by vertical and horizontal movement during the formation of the MCT I. This rock possibly came into the contact with the porphyroblastic gneiss. However, in the field its contact is always with the Pratapnagar quartzite (Fig. 10.8).

With regard to the MCT II (= Jutogh thrust sheet) it may be stated that the lithologies of the MCT IIa ramped for 4 to 5 km and travelled horizontally along a flat surface for about 5 to 12 km. The sheet of MCT IIb ramped from 1 to 4 km and later followed a horizontal trend for about 5 to 11 km. The marble associated with the porphyroblastic gneiss were the main carrier of the MCT IIa and IIb sheets which were affected by horizontal and vertical thrusting movements. Though presently the medium-grade rocks of the MCT IIa and IIb directly rest over the Pratapnagar quartzite yet they initially rode over the migmatized granite gneiss which in turn was the main carrier of the MCT III characterized by ramps and flats.

The restored sections of the Yamnotri Valley revealed that the MCT III ramped from 3 to 4 km. This low-grade thrust sheet horizontally translated for about 7 to 29 km and tectonically piled over the Pratapnagar quartzite.

It can be concluded that on an average the lithologies of the Jankichatti, Phulchatti and Wazri horses, exposed along different sections, during the formation of the MCT II and III witnessed for about -60% of shortening. The lithologies were shortened for about 8.5 km in case of the MCT II and for about 8 km in respect of the MCT III. Therefore, the shortening distance of the lithologies during the MCT II and MCT III was for about 16.5 km or 17 km in the Yamuna Valley region of the Garhwal Himalaya.

References

Agarwal, N. C. and Kumar, G., 1973. Geology of upper Bhagirathi and Yamuna Valleys, Uttarkashi district, Kumaun Himalaya. *Him. Geol.,* **3**: 1-23; Delhi.

Bahuguna, V. K. and Saklani, P. S., 1988. Tectonics of Main Central Thrust in Garhwal Himalaya, U. P. - *J. Geol. Soc. Ind.,* **31** (2): 197-209; Bangalore.

Boyer, S. E. and Elliott, D., 1982. Thrust Systems. - *Bull. Am. Assoc. Pet. Geol.,* **66** (9): 1196-1230; Tulsa.

Butler, R. W. H., 1982. The terminology of structures in thrust belts. *Jour Struct. Geol.,* **4** (3): 239-245.

Cooper, M.A. 1981. The internal geometry of nappes - criteria for models of emplacement. In : 'Thrust and Nappe Tectonics' (Edited by McClay, K.R. and Prince, N.J.). *Special Publs. Geol. Soc. London,* **9**: 335-344.

Dahlstrom, C.D.A. 1969. Balanced cross-sections. - *Can. J., Earth Sci.,* **6**: 743-746; Ottawa.

Dahlstrom, C. D. A. 1970. Structural geology in the eastern margin of the Canadian Rock Mountains. *Bull. Can. Petrol. Geol.,* **18**, 332-406.

Fuchs, G. and Sinha, A. K. 1978. The Tectonics of the Garhwal Kumaun Lesser Himalaya. - *Jb. Geol. B-A,* **121** (2): 219-241, Wien.

Gansser, A., 1964. *Geology of the Himalaya*s - Willey Interscience, 289 p.

Hossack, J.R., 1983.A Corss-Section through the Scandinavian Caledonides constructed with the aid of branch line maps. *Jour. Struct. Geol.,* **5:** 103-111.

Jangpangi, B. S. 1976. Structure and tectonics of the crystalline rocks of Central Himalaya. - *Him. Geol. Semi:* 133-147; New Delhi.

Johnson, M. R. W., 1986. The structural evolution of the Kumaun Lesser Himalaya. – In: Saklani, P. S. (ed.): *Himalayan Thrusts and Associated rocks*: 27-39; New Delhi (Today and Tomorrow's Publ.).

Le Fort, P., 1989. *The Himalayan Orogenic Segment.* – In: Sengor, A. M. C. (ed.): Tectonic evolution of the Tethyan Region: 289-386; *Kluwer Acad. Publ.*

Narayan Das, G. R. Bhatnagar, G. S. Bhattacharya, A. K. and Sharma, Y. C. 1981. Uranium in metamorphites around Brijranigad, Raunsal, Tehri Garhwal (U.P.). In: Saklani, P. S. (ed.): *Metamorphic Tectonites of the Himalaya* 111-121; Today and Tomorrow's Publ.

Pachauri, A. K., 1972. Stratigraphy, correlation and tectonics of the area around Purola, Uttarkashi and Dehradun districts, U. P. - *Him. Geol.,* **2**: 370-387.

Saklani, P.S., 1993. *Geology of the Lower Himalaya (Garhwal),* International Books, Delhi, 246 p.

Saklani, P. S. and Bahuguna, V. K., 1986. Thrust Tectonics of Garhwal Himalaya. In: Saklani, P. S. (ed.): *Himalayan Thrusts and Associated Rocks:* 1-25; Today and Tomorrow's Publ.

Thompson, R.I., 1982. The nature and signifcance of large "blind thrusts" within the Northern, Rocky Mountains of Canada.In : *Thrust and Nappe Tectonics* (Edited by McClay, K.R. and Price, N.J.). *Special Publ. Geol. Soc. London,***9:** 449-452.

Valdiya. K.S., 1978a. Outline of the structure of Kumaun Himalaya. In: Saklani. P.S., (Ed.). *Tectonic Geology of the Himalaya,* 1-14. Today and Tomorrow's Publishers, New Delhi.

Valdiya, K. S., 1978b. Extension and analogues of the Chail Nappe in the Kumaun Himalaya. - *Ind. J. Earth Sci.,* **5**: 1-19.

Valdiya K.S., 1980. *Geology of Kumaun Lesser Himalaya. Wadia Inst. Him. Geol.,* 291 p.

Valdiya K.S., 1981. Tectonics of the Central Sector of the Himalaya. In: Himalaya Geodynamic Evolution. - *Am. Geophys. Union. Geodyn. Ser.* **3**: 87-110.

•□□□•

11

Neotectonics

It is the study of structures and their history since post-Miocene to present (Bates and Jackson, 1980). Some Neotectonic features are preserved in rocks of the Northwestern Himalaya. (*e.g.* Kashmir and Muzaffarabd).

Most recent example of neotectonics is the occurrence of a major earthquake of the intensity of 7.4 on Richter Scale around Muzaffarabad on 8th October, 2005 at 9:20 a.m. (according to Indian time). The epicenter of Muzaffarabad was at about 12 km in depth. At least about 80000 persons were dead. The worst affected areas were Muzaffarabad, Uri, Baramula, Islamabad, Srinagar, Jammu, Kishtwar etc. The cause of this earthquake is ascribed to movement of Indian plate towards the Tibetan plate which reactivated a transverse fault affecting the Main Boundary Thrust.

During Palaeolithic and Neolithic times the Kashmir and Tibetan parts were areas of gentle slope having a mild climate. The continued plate convergence affected by obliquely directed compressions resulted into thrusting, rotation and uplift of the Himalayan rocks of Pirpanjal ranges. (Fig. 11.1,2,3). The tectonic movements caused folding and northward tilting of the Karewa sediments and the Karewa lake was shrunk in its dimension and became a small lake. The formation of deep gorges and the obliteration of the lake took place during the late Pleistocene (~ 200,000 years B.P.).

The Zanskar mountains vary in altitude from 4500 to 6100 m above the sea-level while the Pirpanjal ranges are located from 3600 to 4600 m. The block faulting and erosion of the Pirpanjal

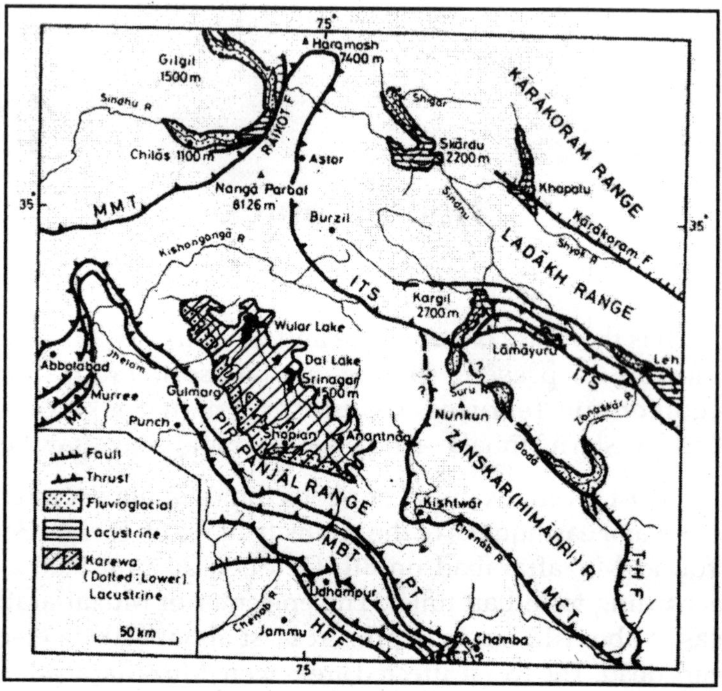

Figure 11.1: Tectonic set up of the northwest Himalaya and location of lacustrine and fluviolacustrine/fluvioglacial deposits of the Quaternary period. (HFF, Himalayan Frontal Fault; MBT, Main Boundary Thrust; MT, Murree Thrust; PT, Panjal (= Chail) Thrust; MCT, Main Central Thrust; T-HF, Trans-Himadri Fault; ITS, Indus-Tsangpo Suture; and MMT, Main Mantle Thrust.) (after Valdiya 1991).

affecting the Main Boundary Thrust, filled the Karewa lake with lacustrine and fluvio-deltaic sediments measuring upto 1300 m in thickness. The age of the Karewas is 4 Ma. These are covered by eolian loess (25 m in thickness) which was deposited at about 25000 years ago (Kusumager et al., in Valdiya, 1991).

The Karewa sediments are characterized by plant fossils. The lake area was at lower altitudes and its climate was temperate type during Pliocene – Pleistocene time. The Indus Valley region had a warm and humid climate which was supported by the fossils of palm (*Sabal major, Livistona, Charophytia gyrogonites* occurring in Hemis conglomerate of Oligocene age (Lakhanpal et al., Gularia et al., in Valdiya, 1991). These plants occur at about 5000 m but

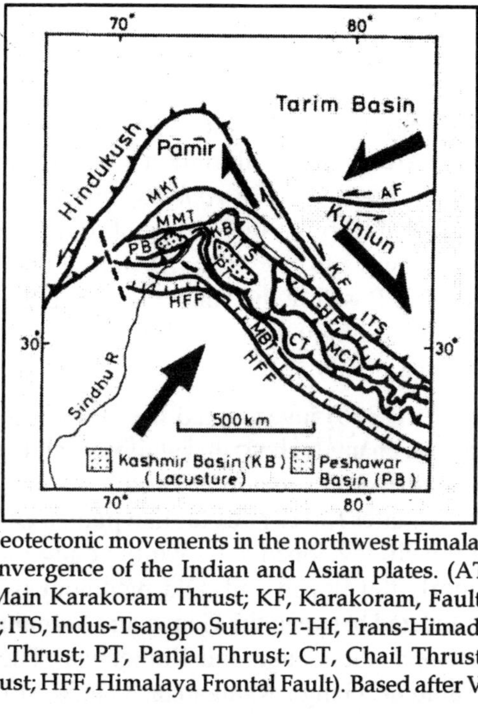

Figure 11.2: Neotectonic movements in the northwest Himalaya - a result of continuing convergence of the Indian and Asian plates. (AT, Altyn Tagh Fault; MKT, Main Karakoram Thrust; KF, Karakoram, Fault; MMT, Main Mantle Thrust; ITS, Indus-Tsangpo Suture; T-Hf, Trans-Himadri Fault; MCT, Main Central Thrust; PT, Panjal Thrust; CT, Chail Thrust; MBT, Main Boundary Thrust; HFF, Himalaya Frontal Fault). Based after Valdiya (1991).

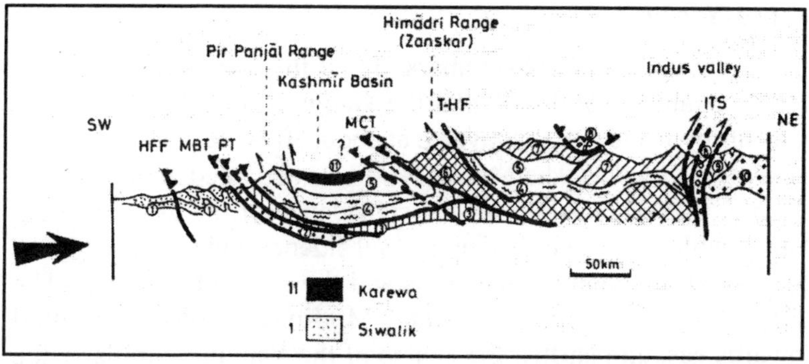

Figure 11.3: Diagrammatic cross-section of the northwest Himalaya explaining neotectonism. (1, Siwalik; 2, Murree; 3, Palaeozoic of Kashmir Basin; 4, Salkhala-Haimanta epimetamorphics; 5, Palaeozoic-Mesozoic succession of Tethys Himalaya; 6, Vaikrita-Zanskar crystallines (basment); 7, Lamayuru-Nindam succession (Triassic-Cretaceous); 8, Ophiolites and ophiolitic melanges of the obduction zone; 9 and 10, Volcanics and plutonics, respectively, of the collision zone. HFF, Himalayan Frontal Fault; MBT, Main Boundary Thrust; PT. Panjal Thrust; T, Tethys Thrust; MCT, Main Central Thrust; T-HF, Trans-Himadri Fault; I-TS, Indus-Tsangpo Suture). Based after Thakur in Valdiya (1991).

are presently found at 2100 m. This suggests about an uplift of 3000 m during Oligocene period. Vertebrate fosils (i.e., *Hippopotamus, Rhinoceras*, *Sivatharium* and *Elephus*) of 0.9 to 0.6 Ma occur in the Karewa sediments. Presently these mammalian animals occur in areas of low altitude having warm and humid climate. Fossils of *ctenodactyles rodents*, *artioductyle* occur in Oligocene molasse of Kargil. Similar type of fossils were found in the Murrees of Jammu and *Stegodon, Elephus* were found in the Siwaliks of Jammu. Possibly the Karewas of Kashmir basin and upper Siwaliks of Jammu had similar heights and climatic conditions (Valdiya, 1991).

Kumar et al., (1996) investigated the Holocene neotectonic effects and climatic changes about the landforms and soil of Ganga Plains. They identified four major geomorphic units, namely, the Piedmonts, Plains, Inter fluves and Aeolian plains. Old Piedmonts are overlain by well developed soils while young piedmonts are overlain by poorly developed soil. Old piedmont is a triangular uplifted block located between the Ganga and Yamuna rivers. It is rectangular between the Ganga and Ramganga. The soil plains include the Yamuna, Ganga Ramganga, Ganga Solani and the Ganga-Yamuna.

Kumar et al., (1996) remarked that the Ganga fault became active at about 2500 B.P. and the Ganga Ramganga block was downthrown and was reactivated at about 500 B.P. which tilted the block towards SW. The Merrut fault was activated at about 5000 BP and was responsible for the subsidence of the upper Ganga-Yamuna block. The vertical uplift / subsidence and tilting of blocks controlled the geomorphic forms and the soil in the region. The upper Ganga-Yamuna block and Modinagar sub-block tilted towards east and shifted the course of the Yamuna. Similarly the Ganga-Ramganga block was also tilted and the Ganga also changed its course (Fig. 11.4). The Yamuna, Merrut, Modinagar and Aligarh faults trend NW-SE oblique to the Himalayan trend and also represent the expression of the basement structures.

The formation of geomorphic features and soil took place in five stages. According to Kumar et.al., (1996); the stage I was 10000-8000 B.P., The climate was dry and cold. The early Holocene vegetation was scanty and the rivers were braided. In stage II (8000-

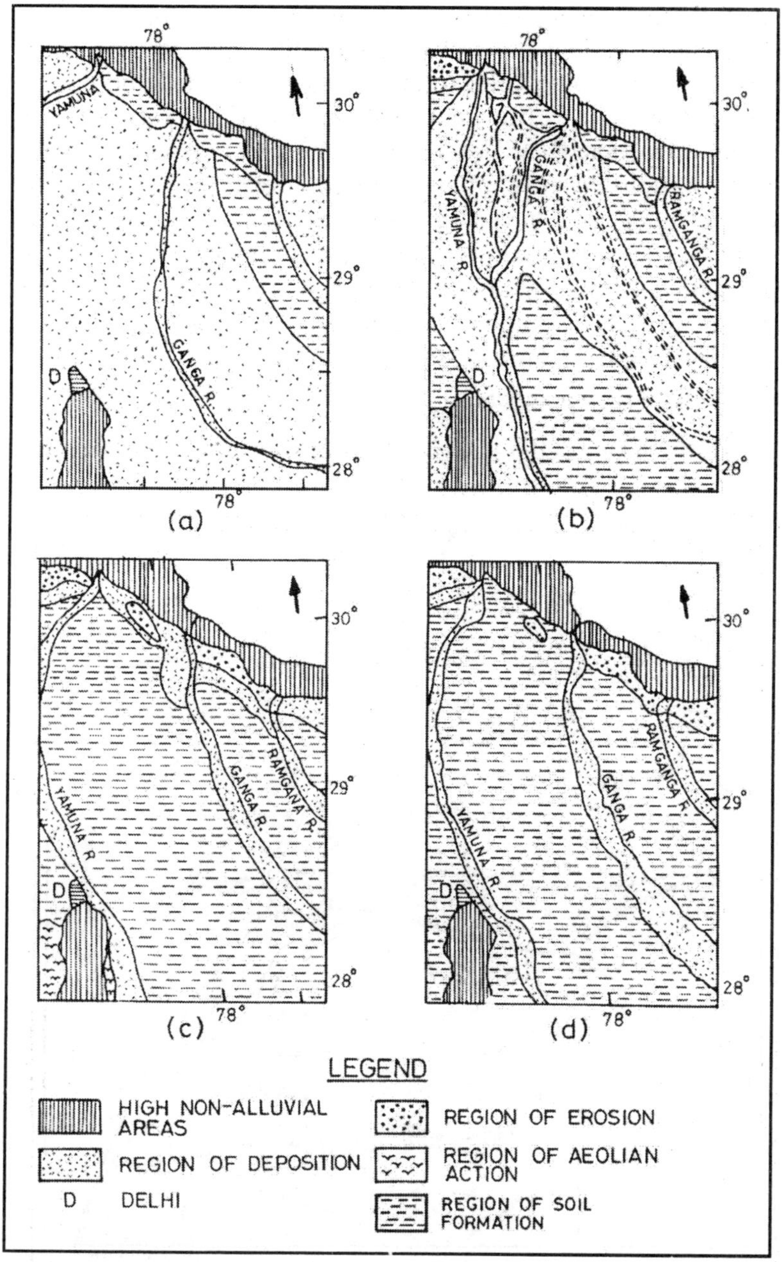

Figure 11.4: Geographic setup in stage I (a), stage III with posible palaeochannels of the Ganga and Yamuna rivers (b) and stages IV (c) and V (d).

500 B.P.) the Ganga, Yamuna and Merrut faults were activated. The climate continued to be cold and dry. During stage III (5000-2500 B.P.), the upper Ganga-Yamuna block subsided and due to this the Yamuna was flowing in Haryana and was a part of Indus system. The stage IV (2500-500 B.P.) the upper Ganga- Yamuna block was uplifted and the Ganga shifted from its present position. At about 500 years to present day the tectonic movements affected the areas. The upper and lower Ganga-Yamuna blocks tilted towards east and the tilting took place due to compression. Presently the depositon of soil is taking place in the flood plains and due to neotectonic movements the Yamuna river is characterized by tightening of the meanders.

Hashmi et al. (1993) described about the Nainital lake and opined that the Nainital massif was uplifted en bloc as a

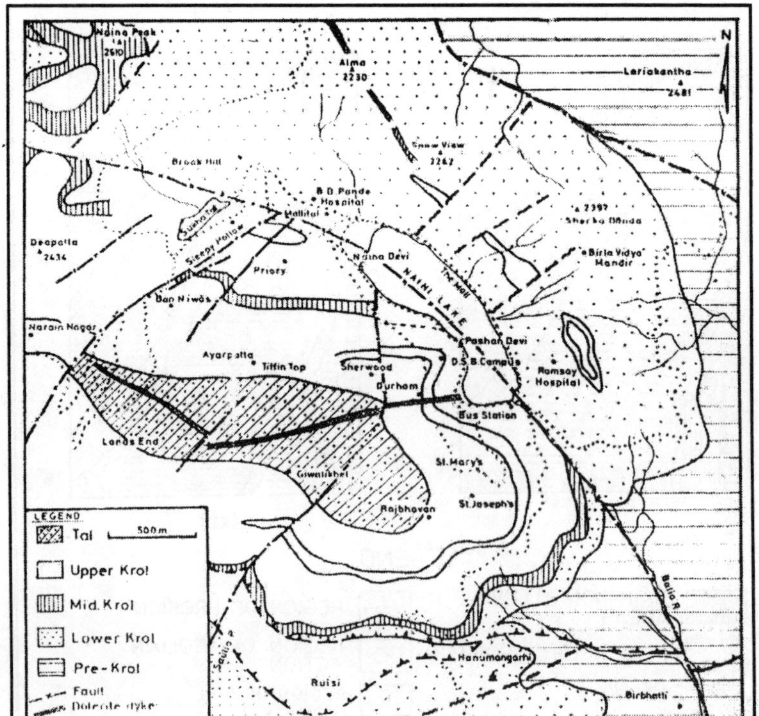

Figure 11.5: Simplified geological map of Nainital traversed by faults (After Valdiya, 1988).

consequence of movement on the Main Boundary Thrust (MBT), triggering extensive mass-movements and gravity gliding on steepened slopes. The lakes of the Nainital district (Khurpa, Naini, Bhim, Naukuchia and Sat "tals") are likewise related to neotectonic movements on the MBT and the genetically related faults (Fig. 11.5).

Vajpai and Kandwal (2005) discussed the synsedimentary tectonics of Delhi. Faridabad and Bulandshahar areas which were

Figure 11.6: (A) Lineaments and their relation with channel pattern of the Yamuna River (Y) in Delhi (D), Faridabad (Fd), Region. S-Sihi, B-Ballabhgarh, H-Hindan River, Ri-Ridge Lineament, Ff-fault, Qtz-quartzite. (After Vajpai and Kandwal, 2005).

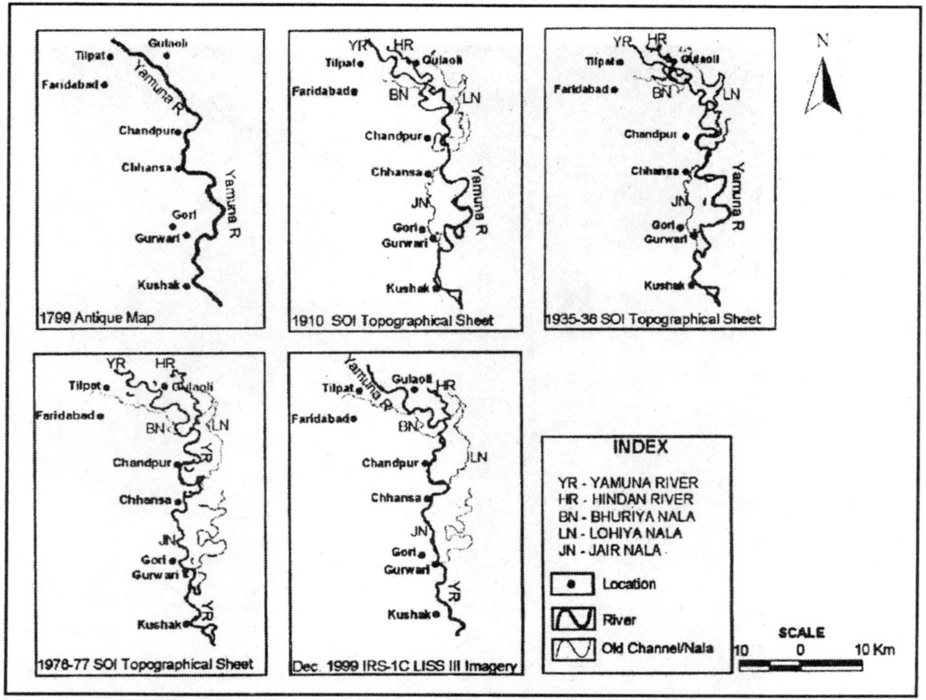

Figure 11.7: Channel changes in the Yamuna River (1799-1999) After Vajpai and Kandwal, 2005.

affected by vertical uplift and subsidence as revealed by the meandering pattern of the Yamuna river during the Quaternary times. The Delhi-Barsana region consists of a graben structure based on Bouguer anomaly data. It is deepest at Hodal (Faridabad). The areas affected by subsidence are characterized by thick disposition of aquifers. They (2005) are of the view that the Yamuna flows along a major fault and where it is criss-crossed by lineaments (faults) the subsidence became more prominent and resulted into sinuous meandering of the river (Fig. 11.7). Changes in the river channels from 1799-1999 are due to neotectonics (Fig. 4).

References

Bates, R.L. and Jackson, J.A., 1980. *Glossary of Geology*, 751 p. *Am. Geol. Inst.*

Hashmi, N.H., Pathak, M.C., Jauhari, P. and Nair, R.R., 1993. Bathymetric Study of the Neotectonic Nainital Lake in outer Kumaun Himalaya, *Jour. Geol. Soc. Ind.,* 91-104.

Kumar, S., Prakash, B., Manchanda, M.L., Singhvi, A.K. and Srivastva, P., 1996, Holocene Landform and Soil evolution of the Western Gangetic Plains : implications of neotectonics and climate, *Z. Geomorph.* N.F., 103, 283-212.

Vajpai, V.N. and Kandwal, K.N., 2005. Synsedimentary Tectonics in the Western Part of the Marginal Gangetic Alluvial Terrain and Its implication for Aquifer Disposition: Case study from Delhi-Faridabad-Bulandhshahr, India special publication of the *Palaeontological society of india,* No. 2, 245-260.

Valdiya, K.S., 1988. *Geology and natural environment of Nainital Hills, Kumaun Himalaya* Gyanodaya, Nainital, 155 p.

Valdiya, K.S., 1991. Quaternary tectonic history of northwest Himalaya. *Current Sci.,* **61** (9&10), 664-668.

• □□□ •

Index

N

O

P

R

S